Health and safety for small construction sites

For people in
BUILDING,
CIVIL ENGINEERING
and
ENGINEERING
CONSTRUCTION

HSE BOOKS

CONTENTS

FOREWORD *V*

INTRODUCTION *I*

1 **SETTING UP THE SITE** *3*

Planning and organising *3*

Common facilities *4*

Control *6*

Monitoring and review *6*

Notifying the site to HSE *6*

Reporting accidents and work-related diseases *7*

2 **WHAT TO LOOK OUT FOR ON SITE** *II*

Working at height *II*
Ladders 12
Tower scaffolds 13
Mobile elevating work platforms 14
General access scaffolds 14
Protection against falling materials 16
Safety harnesses 16
Roof work 17
Industrial roofing 18
Steel erection 19
Formwork and reinforced concrete work 20
Demolition and structural alteration 21

Moving, lifting and handling loads *22*
Manual handling 22
Small lifting equipment 23
Hoists 23
Mobile cranes 25

Groundwork *27*
Excavations 27
Underground services 29

Working in confined spaces *31*

Health hazards *33*
Hazardous substances and processes 33

Asbestos 36
Noise 37
Vibration 38

Protective equipment *39*
Hard hats 39
Footwear 39
Goggles and safety spectacles 40
Outdoor clothing 40
High visibility clothing 40
Gloves 40

Site vehicles and mobile plant *40*

Electricity *42*
Overhead power lines 44

Fire *45*

Work affecting the public *47*

3 **SETTING UP THE COMPANY** *49*

The Health and Safety at Work etc Act 1974 *49*

The construction regulations *49*

The Management of Health and Safety at Work Regulations 1992 *49*

The Construction (Design and Management) Regulations 1994 *52*

Employees' duties *58*

Employment and employees *58*

Inspectors and the law *59*

4 **REFERENCES AND FURTHER INFORMATION** *61*

References *61*

Further information *63*

Area office addresses *64*

iii

FOREWORD

Construction sites can be dangerous places. Far too many people are injured during construction work or have their health damaged by it. This does not just happen on large projects - it happens on small sites too.

This death, injury and ill health causes pain and suffering for those affected, as well as for their friends and family. It also costs money. A recent HSE study[1] showed that 8.5% of tender price was wasted by accidental loss. This was on a site where no one was seriously hurt (suffered a reportable injury) in the incidents. Money was wasted in lost time and broken or damaged material. Most of these losses were uninsurable.

This book tells you how to prevent accidents and injury. Working safely is essential to your business - and the law demands it.

The book is about managing health and safety on sites. This means that first your firm needs to define its overall approach to health and safety, including its health and safety policy and organisation. It then needs to make sure that someone takes responsibility for health and safety. People who run sites need to know enough about health and safety and how to take precautions and maintain them. Employees have to be trained and subcontractors briefed.

The firm also has to ensure that someone is available to provide health and safety advice. Once these arrangements have been made, your firm will be in a position to plan and manage health and safety. The Health and Safety at Work etc Act 1974 and the Management of Health and Safety at Work Regulations 1992 require you to do this at sites where you are working. These legal requirements apply to all work on sites.

Where the Construction (Design and Management) Regulations 1994 (CDM) apply, they provide a framework for an appointed principal contractor to provide overall health and safety management. Other firms working at the site have to co-operate with the principal contractor, but each contractor remains responsible for managing health and safety for their own employees.

This means that on all sites your firm has to manage the health and safety of its own employees and anyone affected by its activities. Where the CDM Regulations apply, your firm may also be acting as the principal contractor. If not, the company will have to co-operate with the principal contractor.

Even if you are genuinely self-employed and work as an independent contractor, you have the same responsibilities as larger firms. You have to ensure your activities do not create risks for others working at the site and co-operate with whoever is managing the site.

Consultants and advisers can be used to help decide how health and safety should be organised, but your firm is responsible for seeing adequate standards are achieved and maintained. Directors and managers of firms also have duties to see that the responsibilities of the firm are discharged.

The CDM Regulations place duties on clients and designers. Clients should assess the competence of those they engage or appoint and provide them with relevant health and safety information. Designers also play a crucial role. Their designs influence health and safety hazards and risks inherent in a project. By understanding this they can, and now should, contribute to reducing health and safety problems within the construction industry.

Following the advice in this book will help you to discharge all these responsibilities and achieve healthy and safe working conditions for yourself, your employees and others affected by your work.

INTRODUCTION

What is this book about?

This book is for smaller construction firms employing up to about 20 people and for those involved with work at smaller construction sites. It explains the key tasks for achieving healthy and safe site conditions by identifying hazards and controlling risks and tells you how to **plan**, **organise**, **control**, **monitor** and **review** health and safety matters throughout the life of the project.

Who should read this book?

The book is for everybody in a small construction firm, especially those managing or supervising site work, including:

(a) directors and partners running a small construction business;

(b) site managers and supervisors running small sites;

(c) managers and supervisors who work on sites run by other companies;

(d) the self-employed.

It may also be useful to clients and designers. It can help clients in the assessment of the competence of their appointees by identifying some of the knowledge required for many small jobs. Designers can also use the book to identify many of the hazards on smaller sites which they need to take into account when preparing their designs.

The advice in the book is for people involved with work on all kinds of construction site, including:

(a) general building and construction work;

(b) refurbishment work;

(c) maintenance and repair work.

(d) engineering construction work;

(e) civil engineering work.

The book is divided into four sections:

1: SETTING UP THE SITE

This tells you about planning, organising, monitoring, controlling and reviewing the job so that you take account of site health and safety from the very beginning of the job through to its completion.

2: WHAT TO LOOK OUT FOR ON SITE

This helps you to identify health and safety hazards found on many sites and gives you advice on how to control the risks that can arise. The book cannot address every hazard, but it focuses on matters which are common causes of death, injury and ill health. Advice is given about protecting the people directly employed to do the work, others working at the site who are affected, visitors to the site and members of the public affected by the work.

Accidents

The most frequent causes of accidental death and injury are:

■ **falls:** People fall because access to and from the workplace is not adequate, or the workplace itself is not safe. Advice is given about safe access to, and work at, height. There are four particularly high risk operations where falls are of major concern: roof work, steel erection, formwork and reinforced concrete and demolition, and these are considered in more detail.

The importance of providing good access, (eg tied ladders) and a safe working place, (eg a platform with toe boards and guard rails) cannot be over-emphasised.

■ **falling material and collapses:** People are struck by material falling from loads being lifted and material which rolls, or is kicked off work platforms; others are struck or buried by falling materials when excavations, buildings or structures collapse. Advice is provided on preventing these types of accidents.

■ **mobile plant:** Construction plant is heavy. It often operates on ground which is muddy and uneven, and where visibility for the driver is poor. People walking on site are injured and killed by moving vehicles, especially reversing vehicles. Others, particularly drivers, are killed and injured by overturning vehicles. This section gives advice on site vehicles and street works where road traffic is a risk to workers and where construction operations are a risk to road traffic and pedestrians.

Ill health

The construction industry has a particularly poor health record. Construction workers are likely to suffer ill health as a result of their employment in the industry after exposure to both hazardous substances and harsh working conditions. Ill health can result from exposure to dusts including asbestos, which causes a range of respiratory diseases and cancer. Exposure to dusts, and also solvents, can cause skin diseases such as dermatitis. Lifting heavy and awkward loads causes back and other injuries. Exposure to high noise levels causes deafness and vibration leads to a range of problems, including vibration white finger. Advice is given on how to control these hazards by eliminating or reducing the problem and where this cannot be done, by providing protective equipment.

3 : SETTING UP THE COMPANY

The law requires you to manage health and safety. This section sets out the law which applies to construction. It gives advice on what this means for the small firm. It tells you what needs to be done to ensure health and safety is dealt with effectively. It also tells you what you can expect others involved in the job to do about health and safety.

4 : REFERENCES AND FURTHER INFORMATION

This section lists sources of further information about site health and safety, including detailed guidance published by the Health and Safety Commission (HSC) and the Health and Safety Executive (HSE).

References are numbered in the text; their full titles are given in this section.

1: SETTING UP THE SITE

The key to achieving healthy and safe site conditions is to ensure that health and safety is planned, organised, controlled, monitored and reviewed.

PLANNING AND ORGANISING

Everyone controlling site work has health and safety responsibilities. Section 3 of the book sets out these responsibilities.

Whether you are the only firm working on a job, a firm running and managing a small job, or a subcontractor working at a site controlled by someone else, you should check that working conditions are healthy and safe before work begins. You should also ensure that you are not going to put others at risk. This requires planning and organisation. Planning has to take account of changes to the site as it develops right through to snagging work and the dismantling of site huts or hoardings at the end of the contract. The basic requirements apply to all sites.

If you have been appointed as a principal contractor under the Construction (Design and Management) Regulations 1994 (CDM) you have further formal responsibilities for securing health and safety on site. These are set out in Section 3 of the book. Whether or not CDM applies, the key tasks to successful health and safety are the same.

Planning the work

Make sure you take account of health and safety information. Get as much information as possible during tendering so that you can allow time and resources to deal with particular problems. Sources of information include:

- the client;

- designers;

- contract documents;

- other contractors at the site;

- specialist contractors and consultants;

- trade and contractor organisations;

- equipment and materials suppliers;

- HSE guidance and British Standards.

Find out about the history of the site and its surroundings. See if there are any unusual features which might affect your work, or how your work affects others, such as:

- asbestos or other contaminants;

- overhead power lines and underground services;

- unusual ground conditions;

- public rights of way across the site;

- nearby schools, footpaths, roads or railways;

- other activities going on at the site.

Where CDM applies, much of this information should be contained within the health and safety plan. If there is a plan make sure you see it and take account of its contents before submitting your tender. Where CDM does not apply you will have to make more enquiries yourself.

When estimating costs and preparing the programme, take account of any particular health and safety hazards associated with the work. Make sure you have made suitable allowances in your price. The job will run more smoothly, efficiently and profitably if you have predicted and planned to control hazards from the outset. It wastes time and money having to stop or re-schedule work to deal with emergencies.

When you buy materials, or hire equipment, the supplier has a duty to provide certain health and safety information. Make sure you get and read the information. For example, you may need to:

- [ ] consider the use of a specialist who is familiar with the necessary precautions;

- [ ] carry out an assessment of the risks arising from the use of a substance and consider using a less harmful substance instead;

☐ provide training on the safe use of the material or equipment.

When preparing your programme consider if there are any operations which will affect the health or safety of others working at the site. For example:

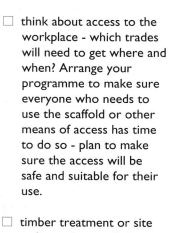

☐ think about access to the workplace - which trades will need to get where and when? Arrange your programme to make sure everyone who needs to use the scaffold or other means of access has time to do so - plan to make sure the access will be safe and suitable for their use.

☐ timber treatment or site radiography usually has to be done when no one else is on site. The site may have to be left vacant for a few days. If you are using a specialist contractor, check the requirements with them and programme the work well in advance.

Adjust your programme to take account of the needs of contractors coming to site.

Discuss proposed working methods with your subcontractors before letting contracts. Find out how they are going to work, what equipment and facilities they are expecting you to provide and the equipment they will bring to the site. Identify any health or safety risks which their operations may create for others working at the site and agree control measures. Obtain health and safety method statements and risk assessments (see Section 3.) to help you. Health and safety method statements with plenty of diagrams are generally more easily understood.

Decide what plant you will require and check it will be suitable.

Consider if workers will need certificates of training achievement, eg for plant operators and scaffolders, before they are allowed to work on the site.

Plan your material deliveries and take account of your storage needs.

Organising the work

Decide on who will supervise your work - check that they are adequately trained and experienced.

When you take on workers, ask about the training they have received, and ask for certificates of training achievement. Get them to demonstrate their knowledge to you or to show you examples of safe working practice before setting them to work.

Make sure that firms coming onto site provide adequate supervision for their workers. Agree what training workers will have received and what you will provide.

See that work methods and safety precautions agreed before work is started are put into practice. Make sure everyone understands how work is to be done and is aware of relevant method statements before work starts.

Find out if any of the work will be further sub-contracted. Make sure that people working for your subcontractors also get the information they require and provide training, supervision, etc as needed.

COMMON FACILITIES

■ **Site access:** There should be safe access onto and around the site for workers and vehicles. Plan how you will keep vehicles clear of pedestrians.

■ **Site boundaries:** Whenever possible, fence off the site. This will protect people (especially children) from site dangers and you from vandalism and theft. For some jobs you will have to share the workplace. Perhaps you will be working in an operating factory or office. Agree which areas are under your control. Agree what fences, barriers, means of separation or permits-to-work are required to keep both construction workers away from hazards created by others and other people away from hazards created by your work. You might need to make site rules (see page 5).

Children may not appreciate the hazards of a construction site - keep sites secure with hoardings.

■ **Site lighting:** Many sites work outside daylight hours and should be well lit. If the building or structure is enclosed, general work areas and access routes, especially stairs, will need good lighting. Where artificial lighting is required, shadows may obscure hazards, so check that enough fill-in lighting is provided.

■ **Falling materials:** Toe boards are required on all raised working platforms. Fans, netting, or hoardings may be needed to protect employees and the public if something goes wrong.

■ **Welfare facilities:** There should generally be toilets, washing facilities, eating facilities and accommodation for ordinary clothing and protective equipment. Make sure someone is responsible for keeping them clean and tidy. Canteens, drying rooms etc should be kept warm and ventilated.

■ **Site tidiness:** Plan how you will keep the site tidy. In particular walkways and stairs should be kept free of tripping hazards such as trailing wires and loose materials. Remove nails from loose timbers to prevent foot and other injuries. Clear paper, timber offcuts and other flammable materials to reduce fire risks.

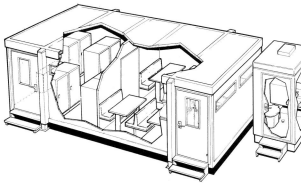

A wide range of portable welfare facilities like these are available. It may be possible when refurbishing buildings to use the facilities already on site.

REMEMBER:
Open-flued gas heaters and gas cooking rings can produce carbon monoxide if not well ventilated. Don't use them in site huts, containers or other enclosed areas unless there is a good supply of fresh air. For further information on welfare facilities, see Construction Information Sheet 18[2].

■ **Fire precautions:** There should be ways of raising the alarm, fighting fires and escaping (see pages 45 to 46).

■ **Storage areas:** Set up storage areas for plant, materials, flammable substances, (eg foam plastics, flammable liquids and gases such as propane) and hazardous substances (eg pesticides and timber treatment chemicals). Do not store materials where they obstruct access routes or where they could interfere with emergency escapes, for example, do not store flammable materials next to fire exits.

■ **Site rules:** Clients may insist on certain safety precautions, especially where their business continues at the premises while you are working. You can help other contractors by making and enforcing site rules. Site rules might cover, for example, safety helmets, safety footwear, site transport, fire prevention, site tidiness, hearing protection or permit-to-work systems. Make sure everybody knows and follows the rules.

Remove nails from old timber as soon as possible to eliminate the risk of foot injury and wounds to other parts of the body.

■ **First aid:** This can save lives, reduce pain and help an injured person to make a quicker recovery. The Health and Safety (First Aid) Regulations 1981 set out the basic requirements. The number of qualified first aiders needed depends on the risk. There should always be on site:

▼ a first aid box with enough equipment to cope with the number of workers on site;

▼ an appointed person who knows how to contact the accident and emergency services; and

▼ information telling workers the name of the appointed person and where to find them. A notice in the site hut is a good way of doing this.

HSE's leaflet *First-aid in your workplace - your questions answered*[3] (IND(G)3L revised) gives more advice to help you.

If you and your workers normally use facilities such as toilets and canteens provided by another contractor, check that they are suitable and properly maintained. Before you start on site make sure that what will be provided is adequate. Otherwise, make arrangements to bring your own facilities to site.

CONTROL

Section 2 of this book identifies a range of hazards common on smaller sites. It also tells you how the risks they give rise to can be controlled. You have to decide which hazards need to be controlled at your sites and what are the best ways of controlling them. Then having decided what needs to be done, you have to make sure it happens in practice. Check that:

■ agreed work methods are put into practice;

■ everyone has the equipment they need; and

■ they are properly trained and competent.

Good site supervision is essential to maintaining healthy and safe site conditions. Make it clear to supervisors exactly what you expect them to do and how you expect them to do it. Consult people working at the site - take their views about health and safety into account.

When people, either your employees, other contractors or visitors, first come to site it is important that they receive information about the site hazards and the steps that have been taken to control the risks. Make sure that the person running the site can be easily identified; if there is a site office, sign it clearly. A site plan showing the office location placed at the site entrance, together with an instruction that all visitors report to the site office, can be helpful.

People who are going to work on the site for the first time should be briefed about risks, welfare facilities and site rules. One way of doing this is by making sure the site supervisor speaks to them before they start work. They might also be given an information sheet or relevant information might be displayed on a notice board prominently placed near to the site entrance. Remember, many people are killed and seriously injured during the first few days that they work at a site.

You can incorporate health and safety checks as part of normal progress along with quality checks you and your supervisors carry out. You may need specific additional checks on higher risk work identified during the planning process.

Carrying out routine checks from time to time reminds everyone that you believe health and safety matters!

MONITORING AND REVIEW

Checking health and safety precautions are being taken is as important as checking progress and quality. Site supervisors need to see that the firm considers the fulfilment of their health and safety responsibilities is an essential part of their job. Section 3 of this book gives more information about this subject.

NOTIFYING THE SITE TO HSE

Sometimes this will be done by someone else, but it may be up to you.

If the construction phase of the work is expected to either:

■ be longer than 30 days; or

■ involve more than 500 person days of construction work,

HSE should be notified in writing before construction work starts. The chart on this page will help you decide if notification is required. The notification should be sent to the nearest HSE office (see page 64 for office addresses). If this has already been done, an extra notification is not required. Ask whoever is co-ordinating work if notification has been sent. A form (10 rev) can be used for notification; a copy of the form can be found on pages 8 and 9. This form can be photocopied and used, or if you prefer, forms are also available from HSE offices. It is not essential that this form is used for notification, but the information required on Form 10 must be provided in writing to HSE. A copy of the notification details should be displayed on site where it can be easily read.

REPORTING ACCIDENTS AND WORK-RELATED DISEASES

The Reporting of Injuries, Diseases and Dangerous Occurrences Regulations 1985 (RIDDOR) require that certain accidents, listed below, that happen on site, have to be reported to HSE (see page 64 for office addresses). You must always report to HSE any of the following accidents which happen to your employees. If you are in control of the site you must also report any accidents which involve a self-employed worker:

☐ serious and fatal accidents must be reported immediately to HSE, normally by telephone;

☐ you must confirm the details in writing within seven days on Form F2508;

☐ less serious injuries, where the injured person is unfit (or unable) to do their normal job for more than three consecutive days must be reported in writing on Form F2508;

☐ if a dangerous occurrence happens on your site, for example, a building, scaffold or falsework collapse, failure of a crane or lifting device or contact with overhead lines, you must immediately, normally by telephone, report it to the nearest HSE office. You must confirm the details in writing within seven days on Form F2508;

☐ if one of your workers suffers from a specified disease associated with their current job, you must report this within seven days.

If a principal contractor has been appointed, contractors should promptly provide them with details of accidents, diseases or dangerous occurrences which are reportable or notifiable under RIDDOR.

HOW TO DECIDE IF YOUR PROJECT HAS TO BE NOTIFIED TO HSE

Will the construction phase be longer than 30 days? → **YES**

NO

WRITTEN NOTIFICATION TO HSE REQUIRED

Will the construction phase involve more than 500 person days of construction work? → **YES**

NO

NOTIFICATION NOT REQUIRED

HSE
Health & Safety
Executive

Notification of project

Note

1. This form can be used to notify any project covered by the Construction (Design and Management) Regulations 1994 which will last longer than 30 days or 500 person days. It can also be used to provide additional details that were not available at the time of initial notification of such projects. (Any day on which construction work is carried out (including holidays and weekends) should be counted, even if the work on that day is of short duration. A person day is one individual, including supervisors and specialists, carrying out construction work for one normal working shift.)

2. The form should be completed and sent to the HSE area office covering the site where construction work is to take place. You should send it as soon as possible after the planning supervisor is appointed to the project.

3. The form can be used by contractors working for domestic clients. In this case only parts 4-8 and 11 need to be filled in.

HSE - For official use only

Client	V	PV	NV	Planning supervisor	V	PV	NV
Focus serial number				Principal contractor	V	PV	NV

1 Is this the initial notification of this project or are you providing additional information that was not previously available

Initial notification ☐   Additional notification ☐

2 **Client:** name, full address, postcode and telephone number *(if more than one client, please attach details on separate sheet)*

Name: Telephone number:

Address:

Postcode:

3 **Planning Supervisor:** name, full address, postcode and telephone number

Name: Telephone number:

Address:

Postcode:

4 **Principal Contractor** *(or contractor when project for a domestic client)* name, full address, postcode and telephone number

Name: Telephone number:

Address:

Postcode:

5 **Address of site:** where construction work is to be carried out

Address:

Postcode

F10 (rev 03.95)

6 Local Authority: name of the local government district council or island council within whose district the operations are to be carried out

7 Please give your estimates on the following: Please indicate if these estimates are original ☐ revised ☐ *(tick relevant box)*

a. The planned date for the commencement of the construction work

b. How long the construction work is expected to take *(in weeks)*

c. The maximum number of people carrying out construction work on site at any one time

d. The number of contractors expected to work on site

8 Construction work: give brief details of the type of construction work that will be carried out

9 Contractors: name, full address and postcode of those who have been chosen to work on the project *(if required continue on a separate sheet) .(Note this information is only required when it is known at the time notification is first made to HSE. An update is not required)*

Declaration of planning supervisor

10 I hereby declare that .... *(name of organisation)* has been appointed as planning supervisor for the project

Signed by or on behalf of the organisation .... *(print name)* ....

Date ....

Declaration of principal contractor

11 I hereby declare that .... *(name of principal contractor)* has been appointed as principal contractor for the project. *(or contractor undertaking project for domestic client)*

Signed by or on behalf of the organisation .... *(print name)* ....

Date ....

See page 64 for list of area offices.

2 WHAT TO LOOK OUT FOR ON SITE

In construction work, most of the hazards are well known. This section identifies many of the hazards you will meet on smaller sites and tells you what to do to control the risks they create.

WORKING AT HEIGHT

Falls are the largest cause of accidental death in construction. Most accidents involving falls could have been prevented if the right equipment had been provided and properly used. If it is possible to fall more than 2 metres (m) from the edge of any working platform, access route or stairway, guard rails or other suitable barriers are needed.

A number of workers are killed or badly injured each year by falls at unprotected openings such as:

- holes in floors;

- between joists;

- lift and service shafts; and

- stair-wells.

Such openings should either be fenced off with secure barriers (eg guard rails and toe boards) or covered over. The cover should either be secured in place or labelled with a warning.

To prevent falls, follow these simple rules:

☐ Decide before you start work at the site what equipment will be suitable for the job and the conditions on site;

☐ Choose a safe method of getting to and from the work area;

☐ Make sure work platforms have guard rails and toe boards;

☐ Make sure that the equipment you need is delivered to the site in good time and that the site has been prepared for it;

☐ Check that the equipment is in good condition and make sure that whoever puts the equipment together is trained and knows what they are doing;

☐ Supervise those you allow to use the

equipment, and make sure they use it properly;

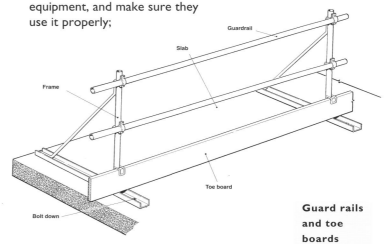

Guardrail

Slab

Frame

Toe board

Bolt down

☐ If equipment is provided for you by another company, check it before you use it;

☐ Find out who to tell if you want any defects remedied or modifications made and tell them;

☐ Only use ladders as workplaces for short periods and then only if it is safe to do so. It is generally safer to use a tower scaffold (see p13) or mobile elevating work platform (MEWP - see p14) even for short-term work;

☐ Treat the use of harnesses and lines to prevent falls as a last resort - they only provide protection for the person using the harness in the event of a fall - they do not prevent the fall itself;

☐ Before any work at height, check that there is adequate clearance from overhead power lines.

☐ There are a number of specialist ways of gaining access to heights, eg rope access, bosuns' chairs and cradles. These need specialist advice and are not covered here.

Guard rails and toe boards should always be securely fixed.

Remember, protection is also required at edges of excavations and where people can fall into water.

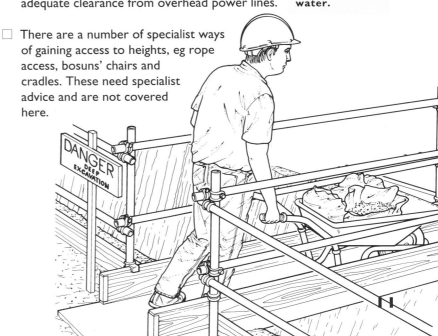

DANGER DEEP EXCAVATION

Ladders

Is the ladder the right equipment for the job?
Ladders are best used as a means of getting to a workplace. They should only be used as a workplace for short-term work.

Ladders are only suitable for light work, so make sure:

☐ the work can be reached without stretching;

☐ the ladder can be fixed to prevent slipping; and

☐ a good hand-hold is available.

However, this kind of work can still be dangerous - many ladder accidents happen during work lasting less than 30 minutes. The longer the ladder, the more problems there are in using it safely. It gets harder to handle, is more difficult to foot effectively and it flexes more in use. Make certain there is no other better means of access before using a ladder longer than 9 m.

Many accidents result from using ladders for a job when a tower scaffold or mobile access platform would have been safer and more efficient.

Are you using the ladder safely?
Carry light tools in a shoulder bag or holster attached to a belt so that you have both hands free for climbing. If you need to raise or lower loads, use a gin wheel or other lifting equipment rather than carrying bulky items up the ladder.

Is the ladder safe?
The ladder needs to be strong enough for the job and in good condition:

☐ check the stiles are not damaged, buckled or warped, no rungs are cracked or missing and safety feet, if fitted, are not missing;

☐ do not use makeshift or home-made ladders or carry out makeshift repairs to a damaged ladder;

This ladder is securely tied to prevent slipping. It is correctly angled and extends above the working platform to allow people to get on and off safely.

☐ do not use painted ladders as the paint may hide faults;

☐ ladders made for DIY use may not survive heavy site use and are best avoided.

Is the ladder secure?
More than half the accidents happen because ladders are not securely placed and fixed. Ladders are only safe when they rest on a firm, level surface. Do not place them on loose bricks or packing.

Also make sure:

☐ the ladder is angled to minimise the risk of slipping outwards; as a rule of thumb the ladder needs to be 'one out for every four up' - see the illustration.

☐ the top of the ladder rests against a solid surface; ladders should not rest against fragile materials;

☐ both feet of the ladder are on a firm footing and that they cannot slip;

☐ if the ladder is more than 3 m long, or used as a way to and from a workplace, it must be fixed at the top, or if this is not possible at its base;

☐ if the ladder cannot be fixed, a second person foots the ladder while it is being used. (This also applies while the ladder is being fixed);

☐ the ladder extends at least 1.05 m above any landing place where people will get on

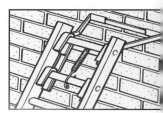

Ladder stays can provide additional security.

This ladder is placed on a board to prevent it sinking into soft ground and tied to stop it slipping.

and off it unless some other adequate hand-hold is available.

Step-ladders

Step-ladders provide a free-standing means of access, but they require careful use. They are not designed for any degree of side loading and are relatively easily overturned. Avoid over-reaching. Workers have been killed getting down from workplaces, such as loft spaces, when they have stepped onto the top step of a step-ladder which has then overturned. Do not use the top step of a step-ladder to work from unless it has specially designed hand-holds.

For further information on ladders, see GS 31[4].

- - - - - - - - - - - - - - - -

Tower scaffolds

Tower scaffolds can be erected quickly and give good, safe access. However, they are involved in numerous accidents each year. These accidents are usually caused either because the tower was not erected properly or because it was not used properly.

If you are going to use a tower scaffold:

- [ ] follow the manufacturer's instructions for erection, use and dismantling. Have a copy of the instruction manual available - if you hire the scaffold the hirer ought to provide you with this information;

- [ ] the tower must be vertical and the legs should rest properly on firm, level ground;

- [ ] lock any wheels and outriggers - base plates provide greater stability if you do not have to move the tower;

- [ ] provide a safe way to get to and from the work platform, eg internal ladders. Climbing up the outside of the tower may pull it over;

- [ ] provide edge protection, (guard rails and toe boards) at platforms from which a person could fall more than 2 m - this applies to the intermediate platforms as well as the working platform;

- [ ] tie the tower rigidly to the structure it is serving if:

  - the tower is sheeted;
  - it is likely to be exposed to strong winds;
  - it is used for carrying out grit blasting or water jetting;
  - heavy materials are lifted up the outside of the tower; or
  - the tower base is too small to ensure stability for the height of the platform.

If ties are needed, check that they are put in place as the scaffold is erected, and checked from time to time (tower scaffolds should be examined in the same way as other scaffolds - see page 15). Also make sure that necessary ties are kept in place as the scaffold is dismantled.

Do not:

- [ ] use a ladder footed on the working platform or apply other horizontal loads which could tilt the tower;

- [ ] overload the working platform.

When moving a mobile tower:

- [ ] check that there are no power lines or overhead obstructions in the way;

- [ ] check that there are no holes or dips in the ground;

- [ ] do not allow people or materials to remain on it as towers tip over very easily when being moved.

For further information on tower scaffolds, see GS 42[5].

Always lock the castors of mobile towers after moving them.

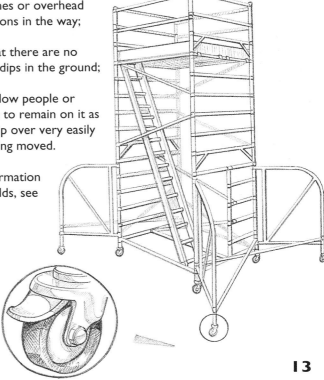

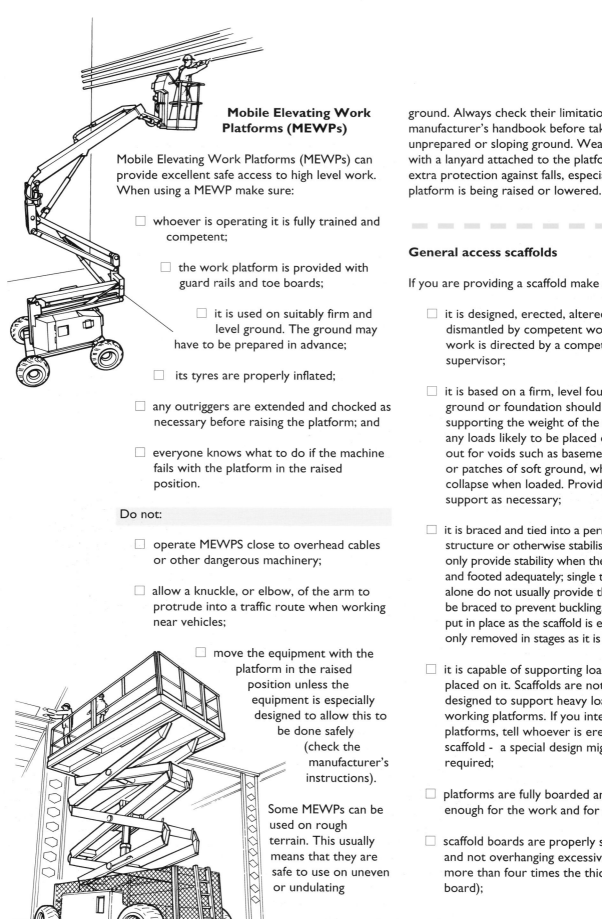

Mobile Elevating Work Platforms (MEWPs)

Mobile Elevating Work Platforms (MEWPs) can provide excellent safe access to high level work. When using a MEWP make sure:

☐ whoever is operating it is fully trained and competent;

☐ the work platform is provided with guard rails and toe boards;

☐ it is used on suitably firm and level ground. The ground may have to be prepared in advance;

☐ its tyres are properly inflated;

☐ any outriggers are extended and chocked as necessary before raising the platform; and

☐ everyone knows what to do if the machine fails with the platform in the raised position.

Do not:

☐ operate MEWPS close to overhead cables or other dangerous machinery;

☐ allow a knuckle, or elbow, of the arm to protrude into a traffic route when working near vehicles;

☐ move the equipment with the platform in the raised position unless the equipment is especially designed to allow this to be done safely (check the manufacturer's instructions).

Some MEWPs can be used on rough terrain. This usually means that they are safe to use on uneven or undulating ground. Always check their limitations in the manufacturer's handbook before taking them onto unprepared or sloping ground. Wearing a harness with a lanyard attached to the platform provides extra protection against falls, especially while the platform is being raised or lowered.

General access scaffolds

If you are providing a scaffold make sure:

☐ it is designed, erected, altered and dismantled by competent workers and the work is directed by a competent supervisor;

☐ it is based on a firm, level foundation. The ground or foundation should be capable of supporting the weight of the scaffold and any loads likely to be placed on it. Watch out for voids such as basements or drains, or patches of soft ground, which could collapse when loaded. Provide extra support as necessary;

☐ it is braced and tied into a permanent structure or otherwise stabilised. Rakers only provide stability when they are braced and footed adequately; single tube rakers alone do not usually provide this and need to be braced to prevent buckling. Ties should be put in place as the scaffold is erected and only removed in stages as it is struck;

☐ it is capable of supporting loads likely to be placed on it. Scaffolds are not usually designed to support heavy loads on their working platforms. If you intend to load-out platforms, tell whoever is erecting the scaffold - a special design might be required;

☐ platforms are fully boarded and are wide enough for the work and for access;

☐ scaffold boards are properly supported and not overhanging excessively (eg no more than four times the thickness of the board);

A wide range of MEWPs is available to provide access for almost any work at height.

☐ there is safe ladder access onto the work platforms;

☐ it is suitable for the task before it is used and inspected at least once a week from then on (or whenever it is substantially altered or adversely affected by, for example, high winds).

The person in control of the scaffold should arrange for it to be inspected weekly. The person doing the inspection needs to understand scaffold safety and record the results of the inspection on Form F91, the scaffold inspection register[6]. Form F91 contains a list of matters to check on the scaffold - use it.

If you are going to use a scaffold provided for you, make sure you don't start work without checking on the scaffold's safety (including a look at the inspection register).

For further information on the construction and safe use of general access scaffolds, see GS 15[7].

Reveal ties only provide reliable support when a reveal pin with a threaded wedge is wound securely into position. As reveal ties are not as secure as other direct fixings they should not make up more than half of the total ties.

All scaffolds, including 'independent' scaffolds should be securely tied, or otherwise supported.

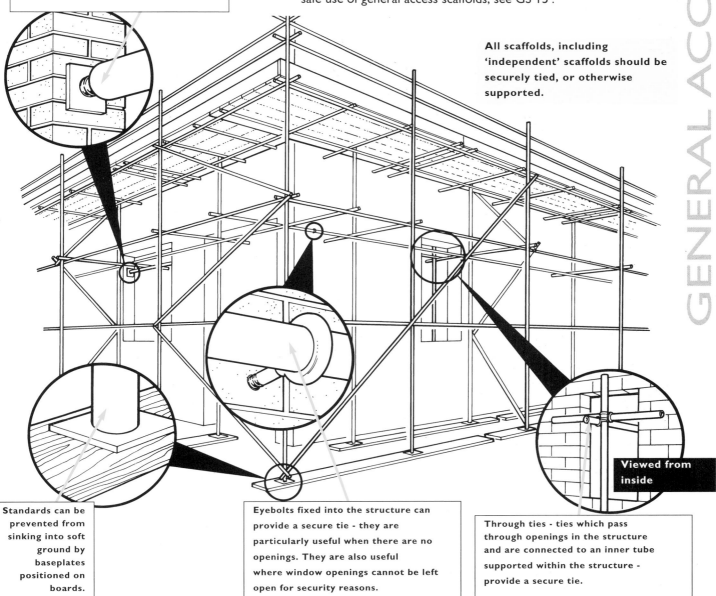

Viewed from inside

Standards can be prevented from sinking into soft ground by baseplates positioned on boards.

Eyebolts fixed into the structure can provide a secure tie - they are particularly useful when there are no openings. They are also useful where window openings cannot be left open for security reasons.

Through ties - ties which pass through openings in the structure and are connected to an inner tube supported within the structure - provide a secure tie.

Protection against falling materials

Toe boards must be provided, as they help to prevent materials rolling, or being kicked, off the edges of scaffold platforms. Brick guards, or similar, will often be needed to provide extra protection against materials falling off the platform. If the scaffold is erected in a public place, scaffold nets and fans may be needed to provide extra protection from falling materials.

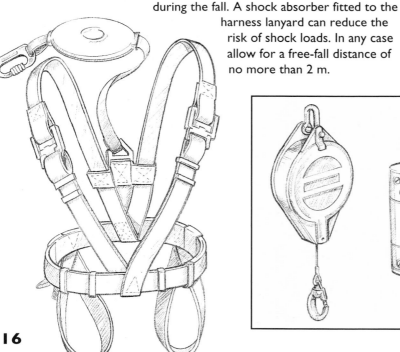

Brickguards should be positioned so that they are prevented from moving outwards by the toe board.

Safety harnesses

The provision of a safe place of work and system of work to prevent falls should always be your first consideration. However, safety harnesses can provide vital protection in some circumstances.

Remember, a harness will not prevent a fall - it can only minimise the risk of injury if there is a fall. The person who falls may be injured by the shock load to the body when the line goes tight or when they strike against parts of the structure during the fall. A shock absorber fitted to the harness lanyard can reduce the risk of shock loads. In any case allow for a free-fall distance of no more than 2 m.

Attach the harness lanyard above the wearer where possible. Additional free movement can be provided by using running lines or inertia reels. Any attachment point must be capable of withstanding the shock load in the event of a fall - you may need expert advice.

Consider how you will recover anyone who does fall.

Make sure everyone knows how to check, wear, adjust and attach their harness before use.

A typical safety harness. Where full harnesses such as this are not used, the maximum possible free-fall distance should be limited to 0.6m.

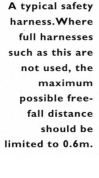

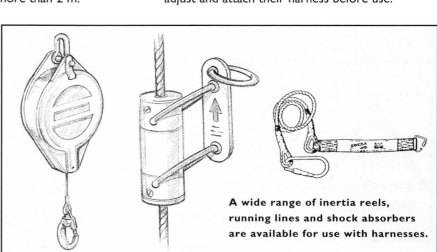

A wide range of inertia reels, running lines and shock absorbers are available for use with harnesses.

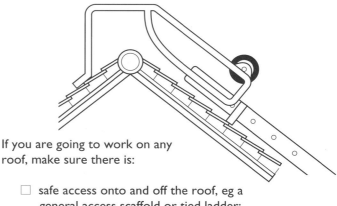

Roof work

Almost one in five workers killed in construction accidents are involved in roof work. Most of these are specialist roofers, but some are simply involved in maintaining and cleaning roofs. Some of these workers die after falling off the edges of flat and sloping roofs.

Many other workers die after falling through fragile materials. Many roof sheets and rooflights are, or can, become fragile. Asbestos cement, fibreglass and plastic generally become more fragile with age. Steel sheets may rust. Sheets on poorly repaired roofs might not be properly supported.

Any of these materials could give way without warning. Do not trust any sheeted roof. Do not stand directly on any of them. On a fragile roof, never try to walk along the line of the roof bolts above the purlins, or along the roof ridge as the sheets can still crack.

Roof openings and rooflights are an extra hazard. Some rooflights are difficult to see in certain light conditions and others may be hidden by bituminous paint.

Openings and lights can be protected by barriers or with a cover either secured in place or labelled with a warning. Labelled covers are generally only suitable where they will not be blown away or otherwise dislodged.

On fragile and most sloping roofs you need purpose-made roof ladders or crawling boards to spread your weight when working on the roof.

DANGER Fragile roof

Guard rails and toe boards erected at the edge or eaves level of a roof are usually needed to stop people and materials from falling off.

If you are going to work on any roof, make sure there is:

☐ safe access onto and off the roof, eg a general access scaffold or tied ladder;

☐ a safe means of moving across the roof, eg roof ladders or crawling boards etc, as appropriate;

☐ a safe means of working on the roof, eg guard rails, a harness or other systems.

Do not throw materials such as old slates, tiles etc from the roof or scaffold - someone may be passing by. Use enclosed debris chutes or lower the debris in containers.

A scaffold platform at eaves level provides safe access for work on a pitched roof.

Industrial roofing

This work involves all the hazards mentioned on page 17. In addition you must consider falls from the 'leading edge'. Leading edges are created as new roof sheets are laid, or old ones are removed. Falls from these edges should be prevented as well as falls from the roof edges and through fragile materials. Fragile and lightweight materials, such as liner trays which will buckle and give way under the weight of a person, can also be a problem at the leading edge and should also be protected.

A movable purlin trolley fitted with guard rails and toe boards.

Work at the leading edge requires careful planning to develop a safe system of work.

Stagings in advance of the leading edge can provide protection; these can be easily moved if proprietary runners are used on the purlins. Guard rails and toe boards can be provided on the stagings.

The correct use of harnesses attached to a suitable fixing can give more protection (see p16) throughout the operation. Running lines and inertia reels may allow easier movement. Nets can provide an alternative means of protection.

When developing a safe system of work also consider:

☐ how the first sheets will be laid - a separate platform may be required - a pack of roof sheets is not a safe working platform;

☐ how sheets will be raised to roof level - decide whether lifting machinery such as an inclined hoist can be used. This eliminates unnecessary risks when placing packs of sheets on the roof supports or when breaking open packs spread over the roof supports;

☐ if the work can be done from separate access equipment such as MEWPs or tower scaffolds. This avoids or minimises the need to work on the roof and so minimises the risk of falls. Using MEWPs (see p14) or tower scaffolds (see p13) may be the safest way to carry out industrial roof work. You should always consider them as a means of access when developing your system of work for a job.

Flat roofs are found on industrial, office and domestic buildings. All roof edges from which people are liable to fall while work is in progress should be protected.

Steel erection

This is high risk work and requires careful planning and execution - it is best left to specialists. The main hazards are:

- falling when working at height;

- erectors being hit or knocked off the steel by moving steel which is being craned into position;

- collapse of the structure before it is fully braced;

- dropping materials onto people working below;

- lifting heavy steel members; and

- cranes overturning.

Before work starts, plan safe working methods:

- ☐ plan for good access onto the site and proper standing areas for delivery vehicles, cranes, mobile work platforms and tower scaffolds;

- ☐ arrange for safe storage of materials;

- ☐ programme work to make sure other trades do not have to work beneath the erectors to avoid the risk of injury from dropped materials;

- ☐ arrange for safe working at height using mobile work platforms, tower scaffolds or another form of independent access if possible;

- ☐ make sure erectors who must work on the frame wear a harness and lanyard which can be connected to the steel work (see page 16);

- ☐ make it clear that workers must not walk on the top flange of steel beams;

- ☐ make sure that all necessary materials (including braces and fixings) are delivered to the site in the correct sequence;

- ☐ check temporary bracing to ensure stability has been provided - consult the designer or a structural engineer;

- ☐ agree a safety method statement and make sure it is followed.

There will always be a high risk of falls if erectors have to work directly off the steel structure. With low rise structures it is generally safer if other access, eg mobile platforms (see MEWPs, p14) or tower scaffolds (see p13) can be used for access for bolting-up and similar operations. It may be necessary to prepare and level the ground before work starts to allow safe use of mobile platforms or tower scaffolds and safe standing for a crane.

Steel erection is a specialist operation and extensive guidance is provided in GS 28 1-4[8,9,10,11]. Similar precautions should be taken when erecting any pre-fabricated structure.

A manlock girder grip device can provide a secure temporary attachment to the steel.

Beams should either be straddled, or if the web is very deep, crabbed.

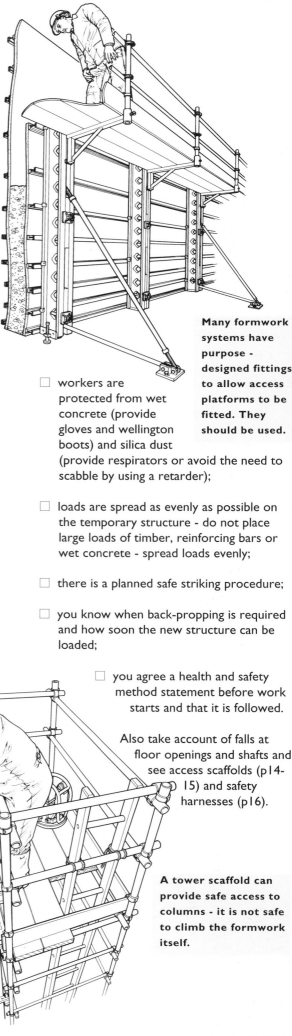

Many formwork systems have purpose - designed fittings to allow access platforms to be fitted. They should be used.

Formwork and reinforced concrete work

The main risks are:

■ people falling during steelfixing and erection of formwork;

■ collapse of the formwork/falsework;

■ materials falling during striking of the formwork;

■ silica dust from scabbling operations;

■ manual handling of shutters, reinforcing bars etc;

■ cement burns from wet cement;

■ arm and back strain for steelfixers.

Make sure:

☐ guard rails to prevent falls are put in place as work progresses;

☐ workers have safe access to the work - it is not safe to stand on primary or other open timbers;

☐ that ladders are tied. Do not climb up vertical sections of reinforcement or the wedges of column forms;

☐ that a ladder or a tower scaffold is used for access;

☐ equipment is in good order before use. Do not use substitutes for the manufacturer's pins of adjustable props;

☐ the formwork, falsework and temporary supports are checked, properly tied, footed, braced and supported before loading, and before pouring walls or columns;

☐ workers are protected from wet concrete (provide gloves and wellington boots) and silica dust (provide respirators or avoid the need to scabble by using a retarder);

☐ loads are spread as evenly as possible on the temporary structure - do not place large loads of timber, reinforcing bars or wet concrete - spread loads evenly;

☐ there is a planned safe striking procedure;

☐ you know when back-propping is required and how soon the new structure can be loaded;

☐ you agree a health and safety method statement before work starts and that it is followed.

Also take account of falls at floor openings and shafts and see access scaffolds (p14-15) and safety harnesses (p16).

Adjustable props. The end plate should be flat and square to the tube. Both tubular sections should be straight and undented. The correct pin should be in place. The collar should be free to move over threads which are clean and lubricated.

A tower scaffold can provide safe access to columns - it is not safe to climb the formwork itself.

20

Demolition and structural alteration

This is high risk work. Falls and premature collapse of structures are the greatest risks. Safe working requires planning. Proposed working methods may be best detailed in a health and safety method statement.

The CDM Regulations (see page 52) apply to all demolition and dismantling work.

It is particularly important that demolition is carried out under the supervision of a competent person.

Before work starts, survey the site and consider:

☐ if a method which keeps people away from the demolition can be used, eg using a long-reach machine or a crane with a ball. Cabs of machines should be protected to safeguard drivers from falling materials;

☐ if the work will make the structure itself, or any nearby buildings or structures unstable. Temporary propping may be needed. You may need the advice of a structural engineer;

☐ if the floors or walls will support the weight of removed material building up on them or the weight of machines, eg skid-steer

loaders used to clear the surcharge. Again, you may need expert advice;

☐ if there are still any live services (eg gas, electricity, water) that need to be dealt with;

☐ if there is any left-over contamination from previous use of the building, eg acids from industrial processes, asbestos on pipework and boilers (see page 36) or micro-biological hazards in old hospitals or medical buildings. Hazardous materials may need to be removed and disposed of safely before demolition starts. Precautions needed for working with hazardous materials are set out on pages 33 to 36.

Keep people away if they are not involved in the work. Create a buffer zone around the work area. Where necessary provide site hoardings. Do not allow materials to fall into any area where people are working or passing through. Fans, or other protection, may be needed to control falling materials.

Fire is also an ever-present risk, so make sure the precautions (see page 45) are in place.

Further guidance can be found in GS 29 Parts 1, 3, 4[12,13,14].

Mechanical demolition allows work to proceed while people are kept away from hazardous areas. The machine driver is protected from falling materials by a cage.

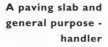

MOVING, LIFTING AND HANDLING LOADS

Every year, a number of construction workers are killed or seriously injured in lifting operations that go wrong; many more end their working lives with permanent back trouble. Some are injured because they lift or carry items which are too heavy or awkward. Many loads would have been better moved using some form of lifting or carrying equipment such as a gin wheel or wheelbarrow hoist. Many people are injured when equipment such as hoists, fork-lifts and cranes are provided, but are not used properly.

Plan for materials handling:

☐ Before the job starts decide what sort of material handling is going to take place and what equipment will be needed;

☐ Avoid double handling - it increases risks and is inefficient;

☐ Make sure that any equipment is delivered to the site in good time and that the site has been prepared for it;

☐ Ensure the equipment is set up and operated only by trained and experienced workers;

☐ Co-ordinate site activities so that those involved in lifting operations do not endanger other workers and vice versa;

☐ Arrange for the equipment to be regularly inspected and where necessary examined and tested by someone who fully understands the safety matters. Make sure records are kept. The Lifting Plant and Equipment (Record of Test and Examination etc) Regulations 1992 give details of what has to be covered in the tests and examinations.

A paving slab and general purpose - handler

creates a risk of injury. If avoidance is not reasonably practicable, employers have to carry out an assessment, reduce the risk of injury as far as reasonably practicable and provide information about the weight of loads.

When you or your workers are involved in manual handling, prevent injury by:

☐ avoiding unnecessary handling;

☐ identifying, before you start work, operations which involve lifting heavy or awkward loads or repetitive lifting operations. Find ways of either avoiding the handling altogether, or using mechanical aids to minimise the amount of manual handling;

☐ sharing heavy or awkward loads which have to be lifted by hand. Remember, while some workers are stronger than others, no one is immune from injury;

☐ positioning loads by machine and planning to reduce the height from which they have to be lifted and the distance over which they have to be carried;

☐ training workers in safe lifting techniques and sensible handling of loads;

☐ not allowing anyone on their own to lift building blocks weighing more than 20 kgs;

☐ ordering bagged materials in small, easily handled sizes where possible.

Anyone injuring their back at work should be encouraged to report the injury, get early medical attention and return only gradually to handling duties.

A manhole cover lifter.

Manual handling

Lifting and moving loads by hand is one of the biggest causes of injury at work. Many manual handling injuries result from repeated operations, but even one bad lift can cause a lifetime of pain and disability. The Manual Handling Operations Regulations 1992 require employers to avoid the need for manual handling if it

For further information on lifting building blocks, see Construction Information Sheet 37[15]. Guidance on the Manual Handling Operations Regulations 1992 is given in booklet L 23[16] and *Getting to grips with manual handling: A short guide for employers* IND(G)143L[17]. Advice on procedures to follow to deal with manual handling problems and a number of practical solutions can be found in *Manual handling: Solutions you can handle* HS(G)115[18].

Small lifting equipment

Gin wheels and similar equipment provide a very convenient means of raising loads. Though simple pieces of equipment, you need to take care when assembling and using them if accidents are going to be avoided. If you are providing a gin wheel or similar for your own use, or for someone else's, make sure:

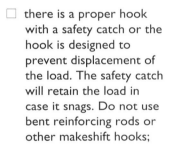

A hook with a safety catch.

- [ ] it has been securely fixed to a secure anchorage, to prevent displacement;

- [ ] there is a proper hook with a safety catch or the hook is designed to prevent displacement of the load. The safety catch will retain the load in case it snags. Do not use bent reinforcing rods or other makeshift hooks;

A hook designed to prevent displacement of the load.

- [ ] there is a safe working platform with guard rails where it is possible to fall when unloading the hook.

Hoists

Select a hoist which is suitable for the site and capable of lifting the loads required.

Set up the controls:

- [ ] so that the hoist can be operated from one position only, eg ground level; and

- [ ] the operator can see all the landing levels from the operating position.

Leaning out to pull in a swivel hoist is hazardous. In this installation a rail which allows access to a wheelbarrow on the hoist platform enables the platform to be pulled in safely.

Also, take precautions against the following:

- ■ **being struck by the platform or other moving parts. Prevent this by:**

  - [ ] enclosing the hoistway at places where people might be struck, eg working platforms or window openings;

  - [ ] providing gates at all landings and at ground level.

- ■ **falling down the hoistway. Prevent this by making sure:**

  - [ ] the hoistway is enclosed where people could fall down it;

  - [ ] the gates at landings are kept closed except during loading and unloading. Gates should be secure and not free to swing into the hoistway;

  - [ ] the edge of the hoist platform is close to the edge of the landing platform so that there is no gap to fall through.

■ **being hit by falling materials. Prevent this by:**

☐ stopping loads falling from the platform, eg make sure wheelbarrows are securely chocked and are not overfilled;

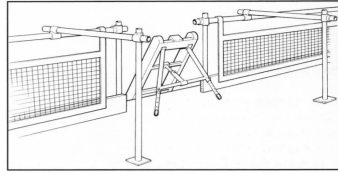

Protection at an upper landing where an inclined hoist passes through edge protection.

☐ not carrying loose loads such as bricks. Put loose loads in proper containers or use a hoist with an enclosed platform;

☐ not overloading the platform. It should be clearly marked with its safe working load;

☐ enclosing the hoistway further.

No one should be allowed to ride on a goods hoist. Put up a notice to say so.

Make sure:

☐ the hoist is erected by trained, experienced people following manufacturers' instructions and properly tied to the supporting structure;

☐ the hoist operator has been trained and is competent;

☐ loads are evenly distributed on the hoist platform;

☐ the hoist is thoroughly examined and tested after erection, substantial alteration or repair at six month intervals. A weekly check should also be carried out and the results recorded in the official register, Form F91[6].

For further information on hoists, see CIS 13[19]. More detail is given in PM 24[20], and 63[21].

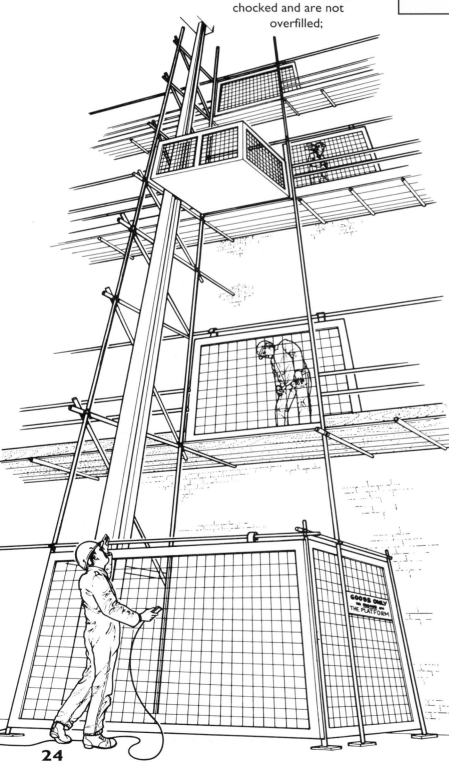

The operator of this hoist has a clear view of each landing. The base of the hoist is protected by a cage and each landing is protected by a sliding gate. The hoist platform has a cage to retain any loose materials.

Mobile cranes

Mobile cranes are a versatile, reliable means of lifting on site. However, it is easy to become complacent about their safe use. Complacency can lead to serious accidents. No lift is small enough to be left to chance. Every lift should be planned and carried out by trained, competent people. If you do not have the expertise, contract out the work to those who do. If you are going to carry out the lift, accidents can be avoided by appointing someone (not the driver) with expertise to take charge. That person will need to plan and manage the lift as follows:

Planning the lift

- [ ] Select the right crane for the job. It will need to be:

  - small enough to get on and off the site and to operate within it; and

  - able to lift the heaviest load at the required radius with capacity to spare. The maximum load a crane can lift decreases the further the load is from the crane, so you may need a crane rated at 20 tonnes to lift a 1 tonne load;

- [ ] Check that certificates are up-to-date - Thorough Examination (within the last 14 months), Test and Thorough Examination of the crane (within the last four years) and Thorough Examination of the lifting gear (within the last six months);

- [ ] Make sure an Automatic Safe Load Indicator (ASLI) is fitted (when the crane is able to lift more than 1 tonne) and is in good working order;

- [ ] The driver should be trained and experienced in the operation of the type of crane being used;

- [ ] Site the crane in a safe place, so that:

  - the driver has a clear view;

  - it is well away from excavations, and overhead power lines;

- it is on level ground which can take its full weight and its load (timber packing may be needed). Check there are no voids such as drains or basements which could suddenly collapse.

Co-ordinating the lift
Make sure:

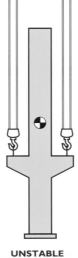

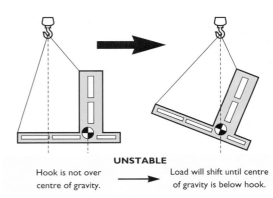

- [ ] the load is properly slung. Chains and slings may be damaged by the load, so packing may be required. The centre of gravity of the load may not be in the middle of the load (this is very common with pieces of

UNSTABLE
Centre of gravity is higher than lift points.

UNSTABLE
Hook is not over centre of gravity.

Load will shift until centre of gravity is below hook.

plant) causing it to shift or slip out of its slings when it is raised. It is important that loads are slung so that they are in balance with their centre of gravity beneath the hook;

- [ ] weekly inspections of the crane are carried out and the results recorded;

- [ ] there is adequate clearance so that people are not struck or trapped by the load, or body of the crane. If traps are inevitable, fence them off;

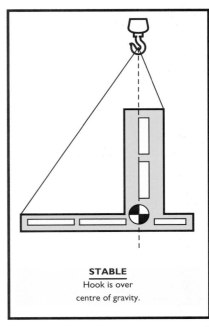

STABLE
Hook is over centre of gravity.

Centre of gravity

Signals to be used by banksmen or signallers.

☐ the crane and the lifting tackle have been checked and maintained as recommended by the manufacturer;

☐ a competent person has been appointed to sling the load. If the driver's view is restricted, a banksman or signaller should be provided.

For further information on the safe use of mobile cranes on site, read BS 7121[22].

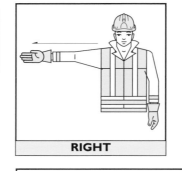

RIGHT

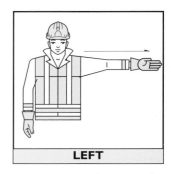

LEFT

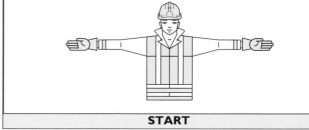

START

DANGER

LOWER

STOP

RAISE

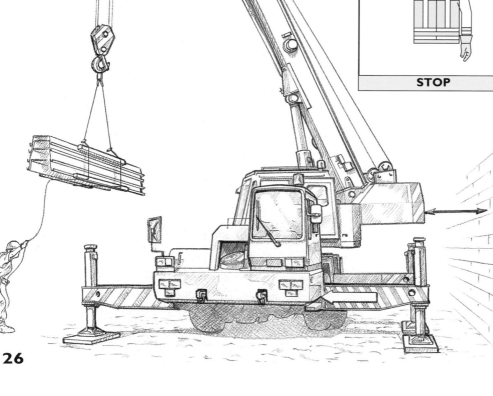

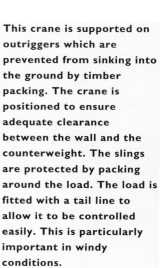

This crane is supported on outriggers which are prevented from sinking into the ground by timber packing. The crane is positioned to ensure adequate clearance between the wall and the counterweight. The slings are protected by packing around the load. The load is fitted with a tail line to allow it to be controlled easily. This is particularly important in windy conditions.

GROUNDWORK

Groundwork can be hazardous. Every year, people are killed or seriously injured while working in excavations. Some are killed or injured when they contact buried underground services. Groundwork has to be properly planned and carried out to prevent accidents.

Excavations

Before digging any trenches, pits or other excavations, decide what temporary support will be required and plan the precautions you are going to take against:

- collapse of the sides of the excavation;

- materials falling onto people working in the excavation;

- people and vehicles falling into the excavation;

- undermining nearby structures etc.

Make sure the equipment and precautions needed such as trench sheets, props, baulks etc are available on site before work starts. If information such as results of soil tests or trial holes is available, it may provide useful details on conditions likely to be found on site which can assist your planning. Put the precautions into practice.

Trench collapse

☐ Prevent trench collapse by battering the sides to a safe angle or supporting them with sheeting or proprietary support systems. Install support without delay as the excavation progresses. Never work ahead of the support. The work should be directed by a competent supervisor and

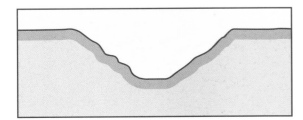

A battered trench.

carried out as far as possible by competent workers with adequate experience. Workers should be given clear instructions.

☐ Someone who fully understands the risks and precautions should inspect the excavation at least once during each day that workers are there. Trenches deeper than 2 m should be inspected before the start of every shift. At least once a week, excavations should be thoroughly examined. They should also be examined after any event (eg a fall of material) that may have affected the temporary support. A record of the examinations should be kept in an official register, form F91[6]. Put right any faults immediately.

Materials falling into excavations

☐ Do not store excavated spoil, plant or materials close to the sides of excavations. Loose material may fall into the excavation and the loading from spoil makes the sides of the trench more likely to collapse. A scaffold board used as a toe board and fixed along the outside of the trench sheets will provide extra protection against loose materials falling into the excavation. Hard hats will protect those working in the excavation from small pieces of materials falling into the excavation or from its sides.

People and vehicles falling into excavations

☐ Prevent workers from falling into unguarded excavations. Edges of excavations more than 2 m deep should be protected by substantial barriers (not barrier tape). All excavations in public places should be suitably barriered off (see diagram on page 30).

☐ Prevent vehicles from falling into excavations by keeping them out of the area - where necessary use baulks or barriers. Baulks and barriers are best painted or marked to make sure they can be seen by drivers. If vehicles have to tip materials into excavations, prevent them from over-running into the excavation by using stop-blocks.

Undermining nearby structures

☐ Make sure your excavation does not undermine the footings of scaffolds or the foundations of nearby buildings or walls. Many walls have very shallow foundations which are easily undermined by even shallow trenches, causing the wall to collapse onto those working in the trench. Before digging starts, decide if extra support for the structure is needed. You may need to survey the foundations and take the advice of a structural engineer.

Other aspects of excavation safety

☐ There should be good ladder access or other means of getting into and out of the excavation safely.

☐ Consider hazardous fumes - do not use petrol or diesel engines in excavations without making arrangements either for the fumes to be ducted safely away or providing for forced ventilation. Do not site petrol or diesel-engined equipment such as generators or compressors near the edge of an excavation; exhaust gases can collect and accumulate in excavations. For information on fumes in confined spaces, including excavations, see page 31.

A range of proprietary trench boxes and hydraulic walings allow trench supports to be put in place without requiring people to enter the excavation.

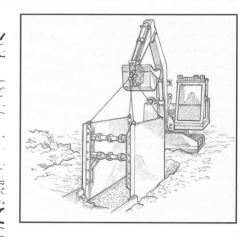

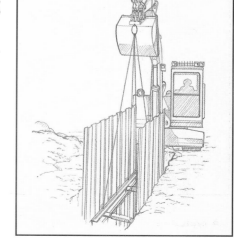

This wall, like many boundary and garden walls, has very shallow foundations. Even a small excavation could undermine the foundations and cause the wall to collapse, crushing anyone nearby.

Underground services

Underground services can be easily damaged during excavation work. If the proper precautions are not taken it is all too easy for workers to hit these services, giving rise to risk:

■ to themselves and anyone nearby from the heat, flame and molten metal given off when an electricity cable is struck, from escaping gas when a gas pipe is hit; or from flooding of the excavation when a water pipe is damaged;

■ from the interruption of services to hospitals, emergency services etc.

All this can be avoided by proper planning and execution. Use cable or pipe plans, cable or pipe locators and safe digging practices.

Use the service plans to see whether the place you intend to dig will involve you working near buried underground services. Look out for signs of services such as manholes, valve covers, streetlights etc. Check for pipes and cables before starting to dig.

Whenever possible keep excavations well away from existing services.

Before digging, make sure:

☐ the person who is going to supervise the digging on site has service plans and is trained in how to use them;

☐ all workers involved in the digging know about safe digging practices and emergency procedures and are properly supervised;

☐ the locator is used to trace as accurately as possible the actual line of any pipe or cable or to confirm that there are no pipes or cables in the way, and the ground has been marked accordingly;

☐ you have devised an emergency plan to deal with damage to cables or pipes. Have a system for notifying the service owner in all cases plus, in the case of gas pipe damage, banning smoking, naked flames and carrying out evacuation whenever necessary (this may include people in nearby properties likely to be affected by leaks).

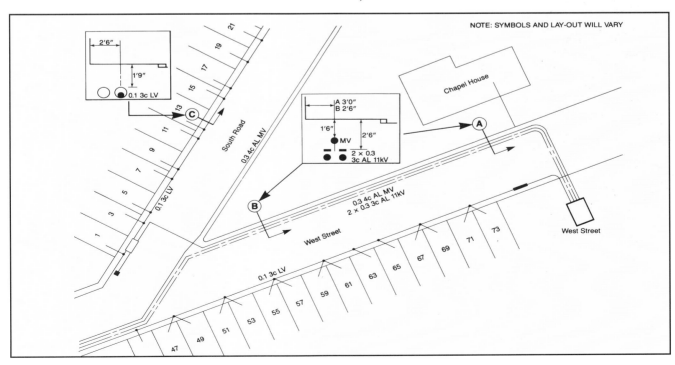

A typical service plan.

Excavate using safe digging practice:

☐ Keep a careful watch for evidence of pipes or cables during digging and repeat checks with the locator. If you meet unidentified services stop work until you know it is safe to proceed;

☐ Hand dig trial holes to confirm the position of the pipes or cables. This is particularly important in the case of plastic pipes which cannot be detected by normal locating equipment;

☐ Hand dig near buried pipes or cables. Use spades and shovels rather than picks and forks which are more likely to pierce cables;

☐ Treat all pipes or cables as 'live' unless you know otherwise. What looks like a rusty pipe may be conduit containing a live cable. Do not break or cut into any service until you are certain of its identity or you know it has been made dead;

☐ Do not use hand-held power tools within 0.5 m of the marked position of an electricity cable. Fit check collars onto the tools so that initial penetration of the surface is restricted;

☐ Do not use a machine to excavate within 0.5 m of a gas pipe;

☐ Support services once they are exposed;

☐ Report any suspected damage to services;

☐ Backfill around pipes or cables with fine material. Backfill which is properly compacted, particularly under cast pipes, prevents settlement which could cause damage at a later date;

☐ Once new services have been laid, update the plans.

For further information on avoiding danger from underground services, read HS(G)47[23].

This excavation is supported by timbering and props. The poling boards extend above the edge of the excavation to act as toe boards and guard rails are provided to prevent falls into the excavation. Safe access is provided by a tied ladder. Exposed services are supported.

WORKING IN CONFINED SPACES

Not knowing the dangers of confined spaces has led to the deaths of many workers. Often the dead include not only those working in the confined space but also those who, not properly equipped, try to rescue them. Work in confined spaces requires skilled and trained people to ensure safety. If part of the project requires confined space working which cannot be avoided, it is often safer to bring in a specialist for the job.

Why is a confined space dangerous?

Air in the confined space is made unbreathable either by poisonous gases and fumes or by lack of oxygen. There is not enough natural ventilation to keep the air fit to breathe. In some cases the gases may be flammable and so there may also be a fire or explosion risk.

Working space may be restricted, bringing workers into close contact with other hazards such as moving machinery, electricity or steam vents and pipes. The entrance to a confined space, for example, a manhole, may make escape or rescue in an emergency more difficult.

How does it get dangerous?

Some confined spaces are naturally dangerous, for example, because of:

- gas build-up in sewers, and manholes and pits connected to them;

- gases leaking into trenches and pits in contaminated land such as old refuse tips, and old gas works;

- rust inside tanks and vessels which eats up the oxygen;

- liquids and slurries which can suddenly fill the space or release gases into it when disturbed;

- chemical reaction between some soils and air causing oxygen depletion or the action of ground water on chalk and limestone producing carbon dioxide.

Some places are made dangerous by vapours from the work done in them. Keep hazards out of

confined spaces. Do not use petrol or diesel engines because exhaust gases are poisonous. Paints, glues etc may give off hazardous vapours.

Ensure the confined space has enough ventilation to make the air fit to breathe. Mechanical ventilation may be needed.

You must have a safe system of work for operations inside confined spaces. Everyone must know and follow the system. A permit-to-work system may be required.

For safe working, first try to find a way of doing the job without going into the confined space. If entry is essential:

- ☐ identify what work must be done in confined spaces and what the hazards are;

- ☐ consider if the space could be altered to make it permanently safe, or the work could be changed to make entry to the hazardous area unnecessary;

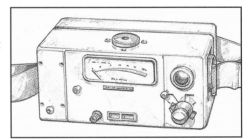

Where necessary use a meter to check air quality before entering a confined space and to monitor the air while work proceeds.

- ☐ make sure workers have been trained in the dangers and precautions, including rescue procedures;

- ☐ make sure the entrance to the space is big enough to allow workers wearing all the necessary equipment to climb in and out easily;

- ☐ before entry, ventilate the space as much as possible, test the air inside the space, and only enter if the test shows it is safe;

- ☐ after entry, continue to test the air for toxic substances, flammable gases and oxygen deficiency as necessary;

- ☐ if there is a flammable risk, the space must be ventilated until it is safe. When selecting equipment, remember heat or sparks from electrical or other equipment could ignite flammable vapours, so air-powered tools may be required. The risk from flammable vapours is very high when work is carried out on the tanks of petrol service stations and similar sites. This is work which may be safer to leave to a specialist contractor;

☐ disturbing deposits and slurries in pipes and tanks may produce extra vapour, giving rise to a greater risk, so clear deposits before entry where possible;

☐ if the air inside the space cannot be made fit to breathe because of a toxic risk or lack of oxygen, workers must wear breathing apparatus;

Here a worker, wearing full breathing apparatus, is wearing a harness with a lanyard connected to a winch so that he can be hauled to the surface in an emergency without others having to enter the manhole to rescue him.

☐ never try to 'sweeten' the air in a confined space with oxygen as this can produce a fire and explosion risk;

☐ workers inside the confined space should wear rescue harnesses, with lifelines attached;

☐ someone should be outside to keep watch and to communicate with anyone inside, raise the alarm in an emergency and take charge of rescue procedures if it becomes necessary. It is essential those outside the space know what to do in an emergency. They need to know how to use breathing apparatus if they are going to carry out a rescue.

For further information on working in confined spaces, read Construction Information Sheet 15[24] and for a more detailed discussion read GS 5[25]. Also see page 33 on COSHH.

Self-contained breathing apparatus (SCBA) - open circuit compressed air type. People wearing this and other types of breathing apparatus, should be trained and competent before beginning work.

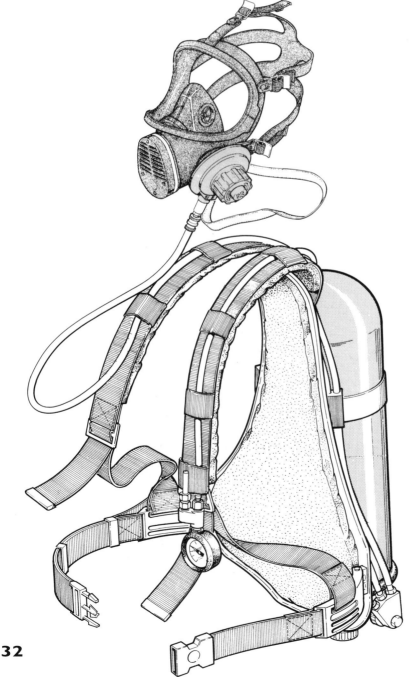

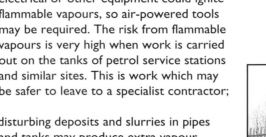

HEALTH HAZARDS

Hazardous substances and processes

You should identify any hazardous substances that you are going to use or processes which produce hazardous dusts or gases. Then you should assess the risks from work which might affect site workers or members of the public. Check with the designer - there may be less hazardous products with a similar performance which could be specified. When carrying out design work, the designer (see page 56) has to consider why hazardous products are needed and where possible, eliminate them.

If you or your workers use hazardous substances or are exposed to them as a result of your work, the Control of Substances Hazardous to Health Regulations (COSHH) 1994 make it a legal duty for you to carry out an assessment to see if they are hazardous and to identify, assess and prevent or control risks to health. There are separate regulations for asbestos and lead - the Control of Asbestos at Work Regulations 1987 (see page 36) and the Control of Lead at Work Regulations 1980. HSE's booklet *Health risk management*[26] gives advice about how to identify and manage health risks.

Identification

People may be exposed to hazardous substances either because they handle or use them directly, eg solvents in glues and paints, or because the work itself gives rise to the hazardous substance, eg scabbling concrete generates silica-containing dust. You should identify and assess both kinds of hazard.

If you use substances which are hazardous as they are supplied to you, the manufacturers and suppliers have a legal duty to provide information. Read the label on the container and/or the safety data sheet. Approach the manufacturer or supplier directly for more information if you think you need it.

In addition some hazardous substances may be on site before you arrive, for example sewer gases (see Working in confined spaces, p31); HSE has also produced guidance about work on contaminated land[27]. You should assess those risks in the same way as for other hazardous substances. Information to help you identify these risks may be available from the client, the design team or the principal contractor (see page 52) or may be found in the pre-tender stage health and safety plan (see page 53).

Assessment

Look at the way people are exposed to the hazardous substance in the particular job you are about to do. You need to decide whether it is likely to harm anyone's health. Harm could be caused by:

- **breathing in fumes, vapours, dust:** Does the manufacturer's information say that there is a risk from inhaling the substance? Are you using large amounts of the substance? Are you working in a way which results in heavy contamination of the air, eg spray application? Are you working in an area which is poorly ventilated, eg a basement?

- **direct contact with skin or eyes:** Does the manufacturer's information say there is a risk from direct contact? How severe is it, eg are strong acids or alkalies being used? Does the method of work make skin contact likely, eg from splashes, when pouring from one container to another, or from the method of application?

If a full assessment has been completed and you do the same work in the same ways under similar circumstances at a number of sites, you do not have to repeat the risk assessment before every job. You should review your assessment from time to time, but every few years is probably enough.

If, however, you carry out many processes which give rise to different hazardous substances in a wide range of circumstances, you might need to do a fresh assessment for each job or set of similar jobs. This makes sure your assessment is relevant to the job you are doing and the circumstances in which you are doing it.

Remember to assess both immediate risks, (eg being overcome by fumes in a confined space) and long-term health risks (materials like cement cause dermatitis; sensitising agents like isocyanates can cause people using them to have severe reactions, even though they may have used the substance many times before).

Prevention

If harm from the substance is likely, the first step to take is to try and avoid it completely by not using it at all. This means either:

- ☐ doing the job in a different way, for instance, instead of using acid or caustic soda to unblock a drain, use drain rods; or

- ☐ using a substitute substance, eg instead of using spirit-based paints, use water-based ones which are generally less hazardous. However, always check you are not simply replacing one hazard with another.

Control

If the substance has to be used because there is no alternative, the next step is to try and control exposure. Some of the ways this could be done include:

- ☐ ensuring good ventilation in the working area by opening doors, windows and skylights. Mechanical ventilation equipment might be needed in some cases;

- ☐ using as little of the hazardous substances as possible - don't take more to the workplace than you need;

- ☐ using cutting and grinding tools fitted with exhaust ventilation or water suppression to control dust;

- ☐ using a roller with a splash guard or applying by

brush, rather than spraying solvent-based materials;

- ☐ transferring liquids with a pump or syphon (not one primed by mouth) rather than by hand. Keep containers closed except when transferring.

Personal protective equipment

If, and only if, you cannot adequately control exposure by any combination of the measures above, you should also provide personal protective equipment (PPE). This might take the form of:

- ▪ **respirators** which can protect against dusts, vapours and gases. Make sure the respirator is of the correct type for the job; dust masks might not protect against vapours or vice versa. If the respirator has replaceable cartridges, make sure the correct type for the job is fitted, that it has not become exhausted or clogged and is still in date (many filters have a limited shelf life). It is essential that respirators fit well around the face. Make sure the user knows how to wear the equipment and check for a good face seal. Respirators do not usually seal well against a beard;

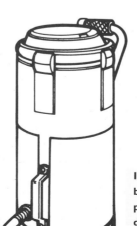

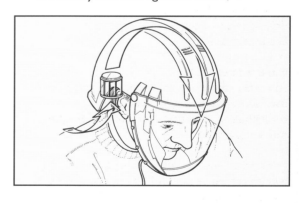

This helmet is fitted with a pump which provides a stream of filtered air to the inside of the visor. These helmets can provide protection for workers with beards as they do not rely on a good seal against the face. The stream of air across the face can also aid comfort in warm environments.

In this chasing operation dust has been reduced by extraction at the tool. This ensures that people working in the area are not exposed to dust. The exposure of the operator is also much reduced, but a dust mask is still required for complete protection.

- **protective clothing**, eg overalls, boots, or gloves. Protection may be needed against corrosive substances;

- **eye protection**, eg goggles, or face visors. Protection of the eyes is important. If the protection needed is against corrosive splashes, visors can protect the whole face.

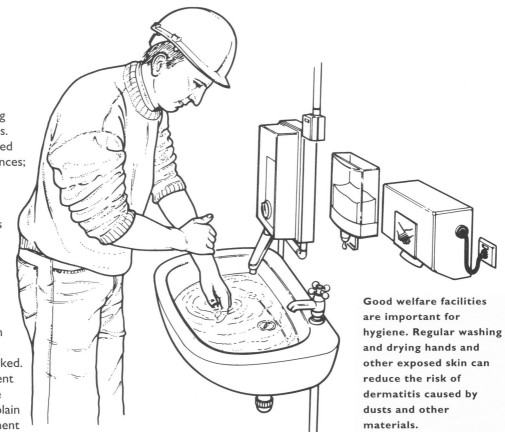

Good welfare facilities are important for hygiene. Regular washing and drying hands and other exposed skin can reduce the risk of dermatitis caused by dusts and other materials.

PPE should be selected with care. Choose good quality equipment which is CE marked. Let the user of the equipment help choose it - they will be more willing to wear it. Explain to the user why the equipment has to be worn, and the hazard(s) it is protecting against. Users need to know how the equipment should be operated and what maintenance checks they should carry out. Supervise the user to make sure it is being used properly. Regularly maintain the equipment and check it for damage. Store it in a dry, clean place and have replacement and spare equipment at hand.

Make sure the PPE does not become a source of contamination, by keeping the inside of dust masks and gloves clean. Store them in a clean box or cupboard - don't leave them lying around in the work area.

For further information on protective clothing and equipment, read L 25[28], IAC (L)16[29] and Construction Information Sheets 28 to 35[30-37].

Personal hygiene
Substances can also be a hazard to health when they are transferred from workers' hands onto food, cigarettes etc and so taken into the body. This can be avoided by good personal hygiene, eg by:

- ☐ washing hands and, where necessary, the face before eating, drinking and smoking and before using the toilet;

- ☐ eating, drinking and smoking only away from the site of exposure.

Make sure those at risk know the hazards. Provide good washing facilities and somewhere clean to eat meals. Good clean welfare facilities can play an important part in protecting the health of everyone involved in the work.

COSHH is explained in *A step-by-step guide to COSHH assessment* HS(G)97[38]. This book also provides a list of many useful references relating to the control of exposure to hazardous substances.

Health surveillance
You can sometimes protect workers' health by checking for early signs of illness. Such surveillance is a legal duty in a restricted range of cases for work involving health risks, eg asbestos. Surveillance may be appropriate, (eg for workers regularly engaged in blast-cleaning, silica-containing stone). The Employment Medical Advisory Service (see page 64 for addresses) can give advice on when and how to carry out health surveillance.

Asbestos

Asbestos has been widely used as lagging on plant and pipework, in insulation products such as fireproofing panels, in asbestos cement roofing materials, and as sprayed coating on structural steelwork to insulate against fire and noise. It is very likely that you will come across asbestos during demolition, or repair and refurbishment work. Often the presence of asbestos will not be obvious, so a careful survey before work begins is always necessary.

If you discover asbestos in the course of work, stop and leave it undisturbed. Protect it from further damage until you have decided how you can proceed in safety. If you are in doubt about the presence of asbestos, assume it is present and work accordingly, or get a specialist analyst to sample the material and test it for you.

Respiratory protection and disposable overalls are needed when working in higher levels of asbestos dust.

All types of asbestos can be dangerous if disturbed. The danger arises when asbestos fibres become airborne. They form a very fine dust which is often invisible. Breathing in asbestos dust can cause serious damage to the lungs and cause cancer. The more asbestos dust inhaled, the greater the risk to health. There is no known cure for asbestos-related diseases.

All work with asbestos and the precautions needed, including respirators, is covered by the Control of Asbestos at Work Regulations 1987. These place a duty on an employer to prevent the exposure of employees to asbestos, or to reduce exposure to the lowest reasonably practicable.

a

WARNING CONTAINS ASBESTOS

Breathing asbestos dust is dangerous to health

Follow safety instructions

Employers should not carry out any work which is liable to expose employees to asbestos unless they have made an adequate assessment of that exposure. It is important to make this assessment even when exposure to asbestos is infrequent and only happens by chance, eg during building refurbishment or repair work such as gas fitting, plumbing or electrical work. The assessment helps employers to decide what precautions need to be taken to protect employees and anyone else who may be affected by the work. Advice for people who may encounter asbestos materials in the course of their work is given in references [39] and [40]

The Regulations are supported by two Approved Codes of Practice (ACoPs), *The control of asbestos at work*[41] and *Work with asbestos insulation, asbestos coating and asbestos insulating board*[42].

> **The Asbestos (Licensing) Regulations 1983 prohibit contractors working on asbestos insulation or asbestos coating unless they have a licence issued by HSE.**

☐ You don't need a licence to work on asbestos cement sheets or most asbestos board, but you must assess the exposure and avoid exposure to airborne dust - respirators may be needed;

☐ Don't break asbestos board or sheet; try to remove it as an undamaged piece. Where you have to work on sheet wet it first if possible;

☐ Use hand tools - drilling and cutting sheet with power tools produces a lot of dust. Use the working methods and precautions described in the asbestos ACoP[41], or other equally safe methods;

☐ If asbestos materials are removed, they must be disposed of safely. Board and sheet materials should usually be wrapped and sealed in polythene sheet and marked to indicate the presence of asbestos. Only specified tips accept asbestos containing waste; check with the local waste disposal authority for details.

The Asbestos (Prohibitions) Regulations 1992 prohibit the import, supply and use of all types of asbestos except chrysotile (white asbestos). Some uses of chrysotile, however, are prohibited. Some prohibitions apply to the supply and use of second-hand products and materials.

New products containing asbestos carry a warning label, as shown on this page.

Noise

Regular exposure to high noise levels causes deafness - the longer you are exposed and the higher the noise level, the greater the degree of deafness which results.

If you or your workers are exposed to noise as a result of your work, you must assess and then prevent or control the risk, providing hearing protection where the risk cannot be eliminated.

Make sure you know which work will involve noisy equipment. Assess how much the noise from this work is going to affect your employees as well as those of other contractors and members of the public.

The manufacturer or supplier of the equipment you use has a legal duty to provide information on the noise it produces. You should get a good idea from the information whether you are likely to have a noise problem. Go back to the manufacturer or supplier if the information is not clear. Where possible choose low noise tools and equipment.

Assessment

- ☐ Look at how you are actually using the equipment on site. Can the person using the equipment talk to someone 2 m away without having to shout to be understood? If they have to shout, the noise from the equipment is probably loud enough to damage their hearing, so you will have to take action.

- ☐ Get the noise levels assessed by someone with the skill and experience to measure noise and who can tell you what you also need to do. In the meantime, offer your workers ear muffs or plugs to wear.

- ☐ Tell all workers exposed above the action levels that there is a risk to their hearing, what you are doing about it and what you expect them to do to minimise the risk.

Prevention

- ☐ Can the job be done in another way which does not involve using noisy equipment? If not, can a quieter item of equipment be used? When buying or hiring equipment, choose the quietest model. Try and carry out the noisy job well away from where other people are working. Move workers not involved out of the noisy area. Erect signs to keep people out of the noisy area.

Control

- ☐ Try and quieten the noise at source, eg fit mufflers to breakers, drills etc. Keep the covers closed on compressors. Most modern compressors are designed to run with all covers closed, even in hot weather. Make sure the silencers on mobile plant are in good condition. Maintain equipment regularly to prevent noise from loose bearings and leaky compressed air hoses and joints.

- ☐ Noise levels can be reduced by making sure the exhausts of compressors, generators and other plant are directed away from work areas. Screens faced with sound-absorbent materials can be placed around plant. Material or spoil heaps can be used to act as noise barriers.

- ☐ If it is not possible to eliminate the noise source or reduce the noise, workers will have to be provided with ear plugs or muffs. Providing hearing protection is not a substitute for noise elimination and control at source.

- ☐ Plugs and muffs should be selected with care, kept in good condition and workers should be trained in their use. Make sure that where muffs or plugs are needed they are actually used. Check that the hearing protection does not interfere with other safety equipment, eg if some ear muffs are difficult to wear with a hard hat, get muffs which fit onto the hat.

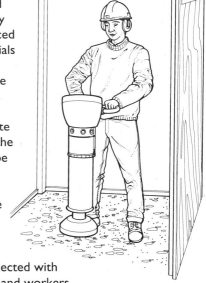

Barriers lined with absorbent material can reduce the spread of noise across the site.

For further information, see *Noise in construction* (revised)[43] and *Dust and noise in the construction process*[44].

Vibration

Many jobs in construction involve the use of hand-held power tools, such as pneumatic breakers and disc grinders. The vibration from such equipment can affect the fingers, hands and arms, and, in the long term, do permanent damage. Parts of the fingers go white and numb and there is a loss of touch.

If you or your workers use hand-held power tools, you should identify, assess, prevent or control the risk from vibration.

You should be able to tell from the manufacturer's or supplier's information whether you are likely to have a vibration problem. Go back to the manufacturer or supplier if the information is not clear. Where possible choose low vibration tools.

Assessment

☐ The information from the manufacturer or supplier, the amount of time the tools are used and discussions with the people using the tools should reveal the tools most likely to present a risk.

Prevention

☐ Can the job be done in another way which does not involve using hand-held power tools (eg by using a hydraulic crusher to break a concrete beam, rather than spending long periods using hand-held breakers)?

Control

☐ Maintain equipment so that it is properly balanced, has no loose or worn out parts, and blades/cutters are sharp etc. Use the power tool and attachment which will do the job properly in the shortest time. Organise work breaks to avoid long periods of uninterrupted vibration exposure.

☐ To protect against vibration, workers should keep their hands warm to get a good flow of blood into the fingers by:

▪ wearing gloves;

▪ having hot food/drinks;

▪ massaging the fingers;

▪ not smoking (this narrows the blood vessels).

For further information on vibration, see *Hand-arm vibration,* HS(G)88[45].

Crushers can be used instead of hand-held breakers. This reduces exposure to vibration, noise and dust; it may also be quicker and cheaper.

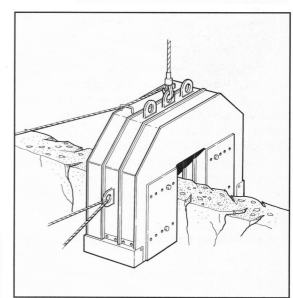

Trenching machines can reduce the need to use vibrating hand tools, although workers may still be exposed to noise and should be protected where necessary.

PROTECTIVE EQUIPMENT

Advice about respirators and other equipment to protect against hazardous substances such as dusts, gases and vapours and against noise and vibration is given in the part on hazardous

substances. This part gives advice about other equipment which may be required to protect against injury.

On almost all sites there is a risk of injury from falling materials and from foot injuries. Minimise these risks by providing toe boards and brickguards at edges of work platforms; keep the site tidy and maintain clear walkways. Deal with the remaining risks by providing suitable hard hats and footwear.

Hard hats

The Construction (Head Protection) Regulations 1989 make specific requirements about hard hats - see HSE's publication[46] for detailed guidance on the Regulations. Hard hats are required where anybody might be struck by falling materials, or where people might hit their heads.

These are just some of the hazards to consider:

- loose material being kicked into an excavation;

- material falling from a scaffold platform;

- material falling off a load being lifted by a crane, or goods hoist or carried on a site dumper or truck;

- a scaffolder dropping a fitting while erecting or dismantling a scaffold.

Decide on which areas of the site hats have to be worn. Tell everyone in the area; if necessary, make site rules.

Provide employees with hard hats. Make sure hats are worn and worn correctly. A wide range of helmets is available. Let your employees try a few and decide which is most suitable for the job and for them. For example, a helmet without a peak, or with a chin strap may be appropriate. Some helmets have extra features, including a sweat band for the forehead and a soft, or webbing harness. Although these helmets are slightly more expensive, they are much more comfortable and therefore more likely to be worn.

Footwear

Is there a risk of injury from either:

- materials being dropped on workers' feet; or

- nails, or other sharp objects, penetrating the sole?

Boots with toe caps and sole-plates may be needed.

39

If it is likely that employees will be working in water or wet concrete, wellington boots will be needed.

Goggles and safety spectacles

These are required to protect against:

- flying objects, eg when shot-firing or using a nail gun. To provide adequate protection they should be shatter-proof - check the manufacturer's specification;

- sparks, eg when disc-cutting;

- ultraviolet radiation from welding; specialist goggles or shields are required.

Outdoor clothing

If your employees regularly work outdoors, you need to provide clothing which is wind and, if necessary, waterproof. There should be facilities for drying wet clothing.

High visibility clothing

Workers acting as banksmen or signallers may need reflective jackets or tabards. People involved in roadworks usually need high visibility clothing.

Gloves

Suitable gloves can protect both against dusts (such as cement), wet concrete and solvents which can cause dermatitis. They can also protect against cuts and splinters when handling bricks, steel and wood.

Here pedestrians are kept clear of vehicles. The roadway is well lit and a banksman is available to guide vehicles out of the site where visibility is restricted.

40

SITE VEHICLES AND MOBILE PLANT

Workers are killed every year on construction sites by moving vehicles or vehicles overturning. Many more are seriously injured in this way. The risks can be reduced if the use of vehicles and mobile plant is properly managed. Plan the site layout to reduce risks:

- provide safe site entry and exit points with adequate turning room and check that vehicle drivers have a good view of any areas where people walk and of other areas where vehicles operate;

- keep pedestrians separate from vehicles by, for example, providing separate site entry and exit points and barriered footways;

- consider a one-way system and avoid the need for vehicles to reverse, wherever possible;

- make use of banksmen or signallers to control high risk situations, eg where vehicles are reversing or visibility is restricted. Banksmen or signallers should be trained and should wear high visibility clothing;

- train the drivers of all vehicles, and make sure visiting drivers are informed about site transport rules;

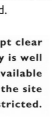

- set out clear routes across the site, avoiding sharp bends, blind corners (suitably placed mirrors aid visibility), narrow gaps, places with limited head room, overhead cables (see page 44), steep gradients, adverse cambers, and shafts and excavations. Provide extra lighting if the area is poorly lit;

- prepare the running surface of temporary roads. Where the site is muddy, skidding and bogging down may be a problem - consider using hard-core or other fill to overcome the problem and repair potholes;

- protect any temporary structures, such as scaffolds or falsework, which might be damaged and made unsafe if struck by a large vehicle;

- provide barriers at excavations and alongside any water if vehicles must pass close by;

- take precautions, eg stop blocks, where vehicles tip materials into excavations;

- do not overload vehicles as they may become unstable, difficult to steer or have their braking efficiency impaired and the load may obstruct the view of the driver;

- make sure loads are securely attached to vehicles and that loose material, such as loose bricks or lumps of clay, cannot fall from lorries or site dumpers and strike pedestrians;

- make sure that mud falling from site vehicles does not cause a hazard on the highway; the local highways authority may be able to provide advice;

- keep plant and vehicles properly maintained. Make sure this is done safely. Never use make-shift jacks to support vehicles while they are being repaired. Never work under unpropped bodies;

- never drain or fill fuel tanks when the equipment is hot, or in a confined space;

- provide refuelling systems that are easy to use, eg hand pumps.

Never work under unpropped bodies.

For further information on site transport safety see *Reversing vehicles* IND(G)148L[47] and *Steering towards safety* IND(G)192L[48].

Sheeting loads prevents loose materials falling from vehicles.

ELECTRICITY

Electrical equipment is used on virtually every site. Everyone is familiar with it, but not all seem to remember that electricity can kill. Electrical systems and equipment must be properly selected, installed, used and maintained.

Electrical equipment used on building sites, particularly power tools and other portable equipment and their leads, faces harsh conditions and rough use. It is likely to be damaged and become dangerous. Modern double insulated tools are well protected, but their leads are still vulnerable to damage.

Where possible, risks should be eliminated. Cordless tools or tools which operate from a 110V supply system which is centre-tapped to earth so that the maximum voltage to earth should not exceed 55V, will effectively eliminate the risk of death and greatly reduce injury in the event of an electrical accident. For other purposes, such as lighting, still lower voltages can be used, and are even safer.

If you need to use mains voltage, the risk of injury is high if equipment, tools, leads etc are damaged, or there is a fault. You will therefore have to take more precautions. Trip devices (such as residual current devices (RCD) rated at 30mA) will be needed to ensure that the current is promptly cut off if contact is made with any live part.

Trip devices must be installed and treated with great care if they are to save life in an accident. They have to be kept free of moisture and dirt and should be checked daily by operating the test button. If mains voltage is to be used, make sure that tools can only be connected to sockets protected by RCDs. If the permanent wiring is being improved as part of the works, you will probably be installing a trip device at the new incoming supply. By installing it at the start of the work you can provide immediate protection. Even so, RCDs cannot give the assurance of safety that cordless equipment or a reduced low voltage (such as 110V) system provides.

Mains equipment is most suitable for use at sites where the work is both indoors where the equipment can be kept dry and where the work will not result in cables etc being damaged by heavy or sharp materials. Where mains leads to sockets may be damaged you should:

☐ position them where they are least likely to be damaged, eg run cables at ceiling height; or

☐ protect them inside non-conducting conduit.

Electrical systems should be regularly checked and maintained. Before using any tool or lead check that:

☐ no bare wires are visible;

☐ the cable covering is not damaged and is free from cuts and abrasions (apart from light scuffing);

☐ the plug is in good condition, ie the casing is not cracked, the pins are not bent or the key way is not blocked with loose material;

☐ there are no taped or other non-standard joints in the cable;

☐ the outer covering (sheath) of the cable is gripped where it enters the plug or the equipment. The coloured insulation of the internal wires should not be visible;

☐ the outer case of the equipment is not damaged or loose and all screws are in place;

☐ there are no overheating or burn marks on the cable or the equipment;

☐ the trip devices (RCDs) are working effectively by pressing the 'test' button every day.

Workers should be trained to report any of these faults immediately and stop using the tool or cable as soon as any damage is seen.

Take damaged equipment out of service as soon as the damage is discovered. Do not carry out makeshift repairs.

In addition to user checks, it is sensible to carry out formal visual inspections of mains voltage systems weekly. Combined inspection and testing should also be carried out by someone trained to do this - see the table for suggested intervals.

With lighting systems you should provide protection for cabling in the same way as you do for tools and make sure bulbs are protected against breakage. If breakage does occur the exposed filaments may present a hazard. Make sure you have a system for checking bulbs to maintain electrical safety and keep the site well lit.

Tools and equipment should be suitable for site conditions. DIY tools, domestic plugs and cables are not designed to stand up to everyday construction work and are best avoided.

If you are going to work in areas where there is a risk of flammable vapours such as at petrol stations or petro-chemical works, your electrical equipment may need to be specially designed to prevent sources of ignition, such as sparks and overheating. Precautions should be covered in the health and safety plan for the project and the operator of the premises should be able to advise you. You may also need to take specialist advice.

SUGGESTED INSPECTION AND TEST FREQUENCIES FOR ELECTRICAL EQUIPMENT ON A CONSTRUCTION SITE

Equipment/ application	Voltage	User check	Formal visual inspection	Combined inspection and test
Battery operated power tools and torches	less than **20 volts**	NO	NO	NO
25V portable hand lamps (confined or damp situations)	**25V** secondary winding from transformer	NO	NO	NO
50V portable hand lamps	secondary winding centre tapped to earth (**25V**)	NO	NO	YEARLY
110V portable hand held tools, extension leads, site lighting, movable wiring systems and associated switchgear	secondary winding centre tapped to earth (**55V**)	WEEKLY	MONTHLY	BEFORE FIRST USE ON SITE AND THEN 3 MONTHLY
240V portable and hand held tools, flood lighting and extension leads	**240V** mains supply through **30mA.** RCD	DAILY/EVERY SHIFT	WEEKLY	BEFORE FIRST USE ON SITE AND THEN MONTHLY
240V fixed (non-movable) equipment	**240V** supply fuses or MCB's	WEEKLY	MONTHLY	BEFORE FIRST USE ON SITE AND THEN 3 MONTHLY
Residual Current Devices (RCD's)		DAILY/EVERY SHIFT	WEEKLY	
Equipment in site offices	**240V** office equipment	MONTHLY	SIX MONTHLY	BEFORE FIRST USE ON SITE AND THEN YEARLY

NOTE: Combined inspection and testing should be carried out:
  (a) where there is reason to suspect the equipment may be faulty, damaged, or contaminated, but this cannot be confirmed by visual inspection; and
  (b) after any repair, modification or similar work on the equipment, when its integrity needs to be established.

Overhead power lines

Contact with overhead electric lines is a regular cause of death and injury. Any work near electric distribution cables or railway power lines must be carefully planned to avoid accidental contact.

The most common operations leading to contact with overhead lines are:

■ handling long scaffold tubes;

■ handling long metal roof sheets;

■ moving long ladders and tower scaffolds;

■ operating cranes and other lifting plant;

■ raising the body or inclined container of tipper lorries; and

■ the use of mobile elevating work platforms (MEWPs).

Where possible all work likely to lead to contact with the overhead line should be done in an area clear of the line itself.

In some cases it may be possible to alter the work to eliminate the risk, for example by reducing the length of scaffold tubes, ladders or roof sheets to ensure that the line cannot be contacted accidentally.

As a general rule no vehicles, plant or equipment should be brought closer than:

■ 15 m of overhead lines suspended from steel towers; or

■ 9 m of overhead lines supported on wooden poles.

In cases where closer approach is likely, it may be necessary to have the lines made dead or to erect barriers to prevent approach to them. If work is to take place close to overhead lines, detailed precautions should be discussed with the owner of the lines (any work next to any railway which is likely to encroach onto railway land should, in any case, be discussed with the railway operator before work begins).

For more details, read HSE's booklet *Avoidance of danger from overhead electric lines* GS 6[49].

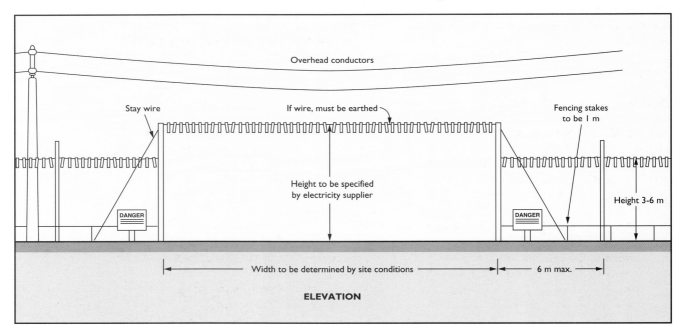

A typical layout to prevent contact with overhead lines. Details of the layout should be discussed with the line owner.

FIRE

Many solids, liquids and gases can catch fire and burn. It only takes a source of ignition, which may be a small flame or an electrical spark, together with air. Any outbreak of fire threatens the health and safety of those on site and will be costly in damage and delay. It can also pose a hazard to surrounding people and properties. Fire can be a particular hazard in refurbishment work when there is a lot of dry timber and at the later stages of jobs where a lot of flammable materials such as carpets and adhesives are in use.

Take precautions to:

- [ ] reduce the possibility of outbreaks of fire;

- [ ] ensure everyone on site can be alerted and can escape from the building in the event of fire;

- [ ] ensure there are efficient arrangements for calling the Fire Brigade in an emergency.

If you or your firm are managing the site, you are responsible for setting up and maintaining the fire precautions. If you or your firm are working at the site, make sure that you know and follow the fire precautions.

Fire prevention

Many fires can be avoided by careful planning and control of the work activities. Good housekeeping and site tidiness are important:

- [ ] Use less easily ignited and fewer flammable materials, eg use water-based or low solvent adhesives and paint;

- [ ] Always keep and carry flammable liquids in suitable closed containers, eg never use open-topped pots or buckets. Keep the quantity at the workplace to a minimum;

- [ ] If work involving the use of highly flammable liquids or solids is being carried out, don't allow other work activities involving potential ignition sources to take place nearby, eg if floor coverings are being laid using solvent-based adhesives, don't allow soldering of pipes at the same time;

- [ ] Minimise the risk of gas leaks and fires involving gas fired plant:

  - close valves on gas cylinders when not in use;

  - regularly check hoses for wear and leaks;

  - prevent oil or grease coming into contact with oxygen cylinder valves;

  - do not leave bitumen boilers unattended when alight;

- [ ] Store flammable solids, liquids and gases safely. Separate them from each other and from oxygen cylinders or oxidising materials. Keep them in a ventilated secure store or an outdoor storage area. Do not store them in or under occupied work areas or where they could obstruct escape routes;

- [ ] Have an extinguisher to hand when doing hot work such as welding or using a disc cutter which produces sparks;

- [ ] Check the site at lunch time and at the end of the day to see that all unattended plant that could cause a fire is turned off. Stop hot working an hour before people go home, as this will allow more time for smouldering fires to be identified;

- [ ] Remove rubbish from the site regularly. Collect highly flammable waste, such as solvent-soaked rags separately in closed fire-resisting containers.

Precautions in case of fire

If a fire should break out, people must be able to escape from it. To achieve this, consider:

- [ ] **means of giving warning:** set up a system to alert people on site; this could be a self-contained fire alarm unit, klaxon, whistle or even word-of-mouth on a small site. The system needs to be distinctive, audible above other noise and recognisable by everyone;

☐ **means of escape:** plan escape routes and ensure they remain available and unobstructed. For elevated work areas within buildings, where possible provide two well separated alternative ways down

to the ground. Signs may be needed if people are not familiar with the escape routes. Make sure that adequate lighting is provided for enclosed escape routes. Protect routes by installing the permanent fire separation and fire doors as soon as possible. Plan for this;

☐ **means of fighting fire:** fire extinguishers should be located at identified fire points around the site. The extinguishers should be appropriate to the nature of the potential fire:

▪ wood, paper and cloth - water extinguisher;

▪ flammable liquids - dry powder or foam extinguisher;

▪ electrical - carbon dioxide (CO_2) extinguisher.

If the building you are working in is occupied, for example, an office, hotel or hospital, make sure your work does not interfere with the building's escape routes, fire separation, alarms, dry risers, or sprinkler systems. Check with the building's occupier or the Fire Brigade.

You should never remove fire doors or leave them propped open. Keep existing dry risers ready for use and install any new ones as soon as possible.

Fire Brigade notification and access

Let the Fire Brigade know where you are, especially for new construction and work above 18 m. Make sure there is adequate access to the site and that neither the access nor hydrants are allowed to become obstructed by building materials.

Further guidance can be found in [50] and [51].

CO$_2$

WATER

FIRE BLANKET

WORK AFFECTING THE PUBLIC

It is not only workers who are at risk from construction work. Members of the public are killed and seriously injured each year; many of these are children.

Accidents often happen when people are walking near a building being built, refurbished or demolished, or walking near street works. Remember, in public areas you need to take account of children, people with prams, the elderly and those with disabilities. Get in touch with the highways authority for advice.

Main points to consider

☐ **Falling materials:** Protect passers-by with brickguards and/or netting on scaffolding, but remember netting only retains light material. Fans and/or covered walkways may be needed where the risk is particularly high. Use plastic sheeting on scaffolds to retain dust, drips and splashes which may occur when cleaning building facades. Make sure the sheets do not make the scaffold unstable; extra scaffold ties may be required.

When using gin wheels or power driven hoists, select a safe place where the public is not at risk. Use hooks with safety catches when lifting (see page 23). Use debris chutes when removing debris into a skip; cover over the skip to stop flying debris and cut down dust.

Remove loose materials and debris from scaffold platforms. Do not stack

Position skips to collect material from chutes or cover their entry. Prevent children using the chute as a slide.

materials on scaffolds unless other protection has been provided (see page 16).

Remove or tie down loose materials and scaffold boards if high winds are possible. Make sure site hoardings will stand up to high winds.

☐ **Works in the roadway or footway:** When you work on the footpath or roadway you may present a hazard to pedestrians and traffic. Road traffic may also present a hazard to the people on site. The Code of Practice *Safety at street works and road works*[52] (available from HMSO) relating to the New Roads and Street Works Act 1991 gives advice about traffic signing, the protection of work areas and pedestrian diversions.

When planning work in streets or similar areas consider:

■ signs for traffic and pedestrians;

■ temporary lighting and traffic controls;

■ cones or other barriers to mark the safety zone;

■ barriers to protect the public. The barrier should alert them to the presence of the site and prevent them from approaching any risks on the site;

■ suitable temporary walking surfaces for pedestrians;

■ storage of materials, for example do not leave paving slabs propped on edge, or pipes loosely stacked in areas where they might be disturbed;

■ how vehicles will manoeuvre into and out of the work area;

Here pedestrians are separated from the works and from traffic.

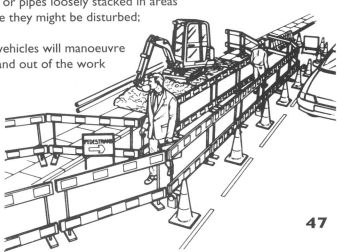

Always ensure that elbows of MEWPs, cranes, excavators, loaders etc cannot swing into the path of vehicles or pedestrians.

■ the provision of high visibility clothing for those working on or next to the roadway;

■ other hazards such as buried services (see page 29) and excavations (see pages 27 and 28) which are dealt with in other sections.

☐ **Tripping hazards:** Excavations in the footpath or roadway should be fenced. Work materials should be stored out of the path of pedestrians and road users. Keep pavements free of tripping hazards and watch out for trailing cables. Make good any damage (temporarily if necessary) as soon as possible.

Extra lighting might be needed at night if there is insufficient street lighting. On some occasions, the pavement will have to be closed to protect the public, eg during pavement work, demolition work, facade cleaning, raising hot asphalt, scaffold erection or dismantling. The area may need to be barriered off and a safe alternative route provided for pedestrians.

☐ **Dusty and hot work:** Fence off hot work such as welding or the use of disc cutters to contain dust and sparks. Fence off boilers which have to be sited in a public place.

☐ **Site visitors:** Make sure site visitors report to the person in charge of the site and know where to go - notices at the site entrance may be needed.

If you are building houses you may well have people wishing to look around, so make sure they are accompanied at all times and given any necessary protective equipment such as helmets or boots. Programme your operations so that work is not in progress on the parts of the site the public visit regularly. Arrange and sign

access routes across the site to keep visitors away from site hazards.

☐ **Site trespassers or intruders:** In most cases you will be responsible for putting up and maintaining the perimeter fence. If you have to alter or take down any safeguards to carry out your work, make sure you put them back as they were when you have finished, or before leaving the site for meal breaks or at the end of the day.

If the site is near a school or on, or near a housing estate, you may find it helpful to contact the head teacher, and residents' association etc to help you discourage children from trespassing. Even though they may be entering the site without authority or trespassing, you must still try and protect them from site dangers; many will be too young to appreciate the risks they are running.

The first precaution is a good perimeter fence. Lock the site gates at night. Take the following steps to reduce the chance of children injuring themselves if they do get onto the site. At the end of the working day:

☐ barrier off or cover over excavations, pits etc. They are a falls risk and may also present a risk of drowning if filled with water;

☐ isolate vehicles and plant; if possible lock them in a compound;

☐ store building materials, such as pipes, manhole rings, cement bags etc so that they cannot topple or roll over;

☐ remove access ladders from excavations and scaffolds; and

☐ lock away hazardous substances.

For further information on preventing accidents to children on construction sites see HSE's publication GS 7[53].

SETTING UP THE COMPANY

All work activities are covered by health and safety law. Not all of it will apply to everything you do, but you need to know the main provisions.

The Health and Safety at Work etc Act 1974

This Act applies to all work activities. It requires employers to ensure so far as reasonably practicable the health and safety of their employees, other people at work and members of the public who may be affected by their work.

You should have a health and safety policy. If you employ five or more people the policy should be in writing. Use the advice in this book to draw up your policy. Keep the policy clear and simple. Make sure everybody in the firm knows about and understands the health and safety systems you have developed. The policy may contain information in the form of health and safety method statements (see page 52).

If you are self-employed you should ensure so far as reasonably practicable your own health and safety and make sure that your work does not put other workers, or members of the public, at risk.

Employees have to co-operate with their employer on health and safety matters and not do anything that puts them or others at risk (see page 58). Employees should be trained and clearly instructed in their duties.

HSE has published guidance on the Act. Booklets HSC 2[54] and HSC 3[55] give outlines of the Act and general advice to employers about it.

The construction regulations

The various construction regulations detail specific requirements relating to construction. They deal with a wide range of health and safety problems including:

- the structure of scaffolds;

- the provision of guard rails to prevent falls;

- the support of excavations to prevent collapse;

- the making of welfare arrangements;

- the use of lifting tackle and lifting equipment, including cranes and hoists.

The Management of Health and Safety at Work Regulations 1992

The Management of Health and Safety at Work Regulations (MHSW Regulations) 1992 apply to everyone at work, regardless of what work it is. Risk assessment is one of their requirements.

Risk assessment

Employers and the self-employed must identify the hazards involved with their work, assess the likelihood of any harm arising and decide on adequate precautions. This process is called risk assessment and is central to all planning for health and safety.

How do I carry out a risk assessment?
A risk assessment can be done in five steps:

**Step 1 -
Looking for the hazards:**

Consider the work, how it will be done, where it is done and what equipment, materials and chemicals are used. The process is like that described for COSHH (see page 33) and noise (see page 37) assessments.

What are the hazards which could cause harm?

Here are some examples which are regular causes of serious and fatal accidents or ill health:

- falling from an open edge or through a fragile material;

- being struck by site vehicles;

- collapse of an excavation or part of a structure;

- work with materials (eg lead, asbestos or solvents) which could be a health problem;

- dust from cutting, grinding, drilling or scabbling.

Many of the common construction site hazards are identified in Section 2.

Step 2 -
Decide who might be harmed and how:

Think about everyone who could be affected by your work: employees, the self-employed, employees of other companies, site visitors and members of the public who may be in the area or outside the site.

Safe working often depends on co-operation between firms. Consider how you need to take them into account in your assessment. Identify problems your work may cause for others at the site, or problems they may cause for you and agree necessary precautions.

Step 3 -
Evaluate the risks and decide on action:

This means asking yourself if somebody is likely to be harmed. Where there is a risk of harm ask yourself:

(i)   **Can the hazard be removed completely?** Can the job be done in another way or by using a different, less hazardous, material? If it can, change the job or process to eliminate the risk.

For example, a house builder's employees were lifting roof trusses into place from the ground by hand. The trusses were heavy and there was a risk that:

▼   *the workers would suffer strain injuries; and*

▼   *the trusses would be dropped and injure someone.*

By using a crane these risks were eliminated.

(ii)   **If the risk cannot be eliminated, can it be controlled?** Applying the advice and guidance given in Section 2 will help you here.

For example, while it may be necessary to apply a solvent-based material, the exposure of workers to hazardous vapours may be reduced by applying it by brush or roller rather than by spraying.

If you have not applied the precautions from Section 2, is what you are doing providing an equivalent or better standard of protection? If it is not, you need to do more.

(iii)   **Can protective measures be taken which will protect the whole workforce?**

For example, to prevent falls, guard rails at edges provide safety for everyone in the area. Secured harnesses only provide safety for those wearing them and are a second best option.

Step 4 -
Record the findings:

Employers with five or more employees should record the significant findings of their assessment. Employers should pass on information about significant risks and the steps they have taken to control the risks, even when they employ less than five people.

Step 5 -
Review the findings:

Reviews are important. They take account of unusual conditions on some sites and changes in the way the job is done. They also allow you to learn from experience. You do not have to produce a new assessment for every job, but if there are major changes you may have to do so. In other cases only the principal contractor is in a position to do a full assessment.

For example, it may be the potential interaction of two or more contractors that leads to increased risk.

In such cases the principal contractor should take the lead (see page 57).

In some cases, eg for steel and formwork erection, or for demolition work, you may need to draw up a safety method statement to take account of risks identified by your assessment (see page 52).

The MHSW Regulations also require employers to have arrangements to:

- plan;

- organise;

- control;

- monitor and review

their work and to:

- have access to competent health and safety advice;

- have arrangements to deal with serious and imminent danger;

- provide health and safety information and training to employees; and

- co-operate in health and safety matters with others who share the workplace.

Health and safety training and advice

Somebody needs to be responsible for the firm's health and safety functions. This will make sure health and safety is not missed or ignored and allows an expertise to be built up in the firm. The people or person responsible may need additional training in health and safety to meet this responsibility properly. If you don't do the job yourself, whoever you appoint must have your full backing.

If you think you need extra advice, you can get it from a number of sources. There are other commercial sources of training which can be found listed in local telephone directories. You can, if you want, arrange your own training in-house. You need to keep your company up-to-date with developments in health and safety, such as new legal requirements.

If there isn't adequate expertise in the company you can get help from the following:

- the Construction Industry Training Board (CITB);

- employers and trade organisations such as the

Building Employers Confederation (BEC) or Federation of Master Builders (FMB);

- Training and Enterprise Councils and Local Enterprise Companies;

- local health and safety groups;

- insurance companies;

- suppliers who must provide instructions on using machines, tools, chemicals etc and product data sheets. Also, containers often have helpful labels;

- safety magazines which have useful articles and advertise safety products and services;

- the British Safety Council (BSC), the Royal Society for the Prevention of Accidents (RoSPA) and many other independent companies and consultants run training courses - look in the telephone directory;

- HSE publishes a newsletter about new HSE publications, changes in the law and similar items of interest and a twice yearly news-sheet *Site Safe News*. If you wish to receive copies the address to contact is at the back of this book (see page 63).

HSE booklet IND(G)133L *Selecting a health and safety consultancy*[56] gives further advice on this subject.

Site supervisors (eg agents, chargehands)

Part of the way you control risks and what happens on site will be through your supervisors and foremen. They direct the way in which work is carried out on site. This means they can and should ensure that work is safe. They also have an important role in passing on training and information to workers on site, eg by giving 'toolbox' talks. However, they cannot do this properly unless they are trained in safe and healthy working practices (see Section 2).

Trades (eg bricklayers, carpenters, scaffolders)

Workers must be trained in safe working practices. You cannot rely on employees picking up safety on the job from their workmates - they might simply be learning someone else's bad habits. You need to be sure of their abilities before setting them to work and to provide necessary training where it is required.

Monitoring health and safety

With any business activity checks need to be made from time to time to make sure that what should be happening is actually being carried out in practice; this includes health and safety. You need to make sure that everyone is fulfilling their duties. If you nominate a supervisor or use the services of a safety adviser, to visit sites and review safety, do they report problems to the site manager and to you? Are matters put right? Do the same problems keep recurring? If there are problems, find out why. Keeping a record of accidents, illnesses and treatments given by first aiders will help you to identify trends. You may need to issue new instructions or provide extra training.

Act before there is an accident or someone's health is damaged. If an accident happens find out what happened and why. Minor accidents and 'near misses' can give an early warning of more serious problems. Consider whether the accident would have happened if your work had been better planned or managed or your employees had been better trained. Could site or company rules have been clearer or could plant and equipment have been better maintained? Don't just put the blame on human error or other people without thinking why the error was made.

Health and safety method statements

In many instances, for higher risk work (eg steel and formwork erection, demolition or the use of hazardous substances) it may be appropriate to draw together the information you have compiled into a health and safety method statement. Information will include the various hazards and the ways in which you will control those hazards for any particular job. Health and safety method statements are not required by law, but they have proved to be an effective and practical management tool.

The health and safety method statement takes into account the conclusions of your assessments under the MHSW Regulations as well as any assessments required by the Control of Substances Hazardous to Health Regulations 1994, the Manual Handling Operations Regulations 1992 etc. It also takes account of your company's health and safety organisation and training procedures and may include arrangements to deal with serious or imminent danger. It can also form part of your company's health and safety policy.

The health and safety method statement sets out how you intend to carry out a job or process, including all the control measures you intend to apply. This will help you to plan the job and identify the health and safety resources you will need for it. It can also provide information for other contractors working at the site about any effects your work will have on them and help the principal contractor to develop an overall health and safety plan for the construction phase of a project (see page 54).

If you repeat a similar operation the statement will be similar from job to job. Where circumstances change markedly, for example with demolition, you may need to revise your statement for each job.

The health and safety method statement is an effective way of providing information to your employees about how you expect the work to be done and the precautions you intend should be taken. The most effective health and safety method statements often have a number of diagrams to make it clear how work should be carried out. Checking that the working methods set out in the statement are actually put into practice on site can also be a useful monitoring tool.

Use the information from your monitoring when you review your risk assessments to help you decide if you are applying adequate precautions.

The Construction (Design and Management) Regulations 1994

What are these Regulations?

The Construction (Design and Management) Regulations 1994 (CDM) are new regulations. They are needed because the construction industry has an unacceptably high rate of fatal and serious accidents and a poor record of ill health.

The Regulations also implement part of an EC Directive.

Many cases of injury and ill health could be prevented by better management of construction work. Other cases of injury and ill health occur because designers do not consider adequately the safety or health of those involved in the construction process. Designers can eliminate hazards before they ever arise on site. In many cases designers choose what risks will occur; contractors then have to manage those risks.

The CDM Regulations require that health and safety is taken into account and managed throughout all stages of a project, from conception, design and planning through to site works and subsequent maintenance and repair of the structure.

The CDM Regulations affect everyone who takes part in the construction process: the client, the designers and contractors. The Regulations introduce two new roles: the **planning supervisor** and the **principal contractor**. The Regulations also introduce **health and safety plans** and the **health and safety file**. You need to be sure your health and safety policies and procedures take account of the requirements of the CDM Regulations.

When do the Regulations apply?

The Regulations apply to most common building, civil engineering and engineering construction work. The Regulations do not apply to very minor construction work if the local authority is the enforcing authority for health and safety purposes. This means that where work is **not notifiable** (see page 57) **and** is either:

- carried out inside offices, shops and similar premises where the construction work is done without interrupting the normal activities in the premises and without separating the construction activities from the other activities; or

- the maintenance or removal of insulation on pipes, boilers or other parts of heating or water systems,

it is not subject to the CDM Regulations.

Apart from this exception, the CDM Regulations apply to **all design work** carried out for construction purposes.

The CDM Regulations apply to **all demolition and dismantling work**. They also apply to most other construction work unless:

- the work will last 30 days or less; **and** involve less than five people on site at any one time; or

- the work is being done for a domestic client (that is someone who lives, or will live, in the premises where the work is being done). In this case only the duty to notify HSE (see page 7) and the designer duties apply.

However, in some instances domestic clients may enter into an arrangement with a developer who carries on a trade, business or other activity. For example, a developer may sell domestic premises before the project is complete. The domestic client then owns the incomplete property, but the developer still arranges for the construction work to be carried out. In this case the CDM requirements apply to the developer.

Use the chart on page 55 to check if the CDM Regulations apply to the work you are doing.

REMEMBER:
Even if the CDM Regulations do not apply, the other regulations continue to do so.

What is the health and safety plan?

The health and safety plan develops with the project and has at least two clear phases (the first associated with design and planning of the project before tendering or contractor selection, the second is associated with the construction phase). The planning supervisor is responsible for seeing that the plan is started.

The pre-tender stage health and safety plan includes:

- a general description of the work;

- details of project timescales;

- details of health and safety risks as far as they are known. This includes information provided by designers about risks they were unable to eliminate and assumptions in broad terms they have made about precautions which will be taken;

- information required by possible principal contractors to allow them to identify the health and safety competences and resources they will need for the project;

- information on which to base a construction phase health and safety plan.

  The pre-tender stage health and safety plan needs to be available to possible principal contractors at the start of selection or tendering procedures. It informs them of the health, safety and welfare matters they need to take into account when planning for site work.

For the construction phase the principal contractor develops a health and safety plan which:

- [ ] sets out how health and safety, and particularly the risks highlighted in the pre-tender stage plan, will be managed during the construction phase, (including details of how information and instructions will be passed to contractors and how their activities will be co-ordinated);

- [ ] includes contractors' risk assessments and health and safety method statements for high risk activities;

- [ ] contains enough information about welfare arrangements to allow contractors on the project to understand how they can comply with welfare requirements;

- [ ] sets out common arrangements (eg on welfare, site hoardings and emergency procedures);

- [ ] details how contractors, material suppliers and plant and equipment supplied for common use, will be selected;

- [ ] sets out how the views of workers and their representatives on health and safety issues associated with the project will be co-ordinated;

- [ ] includes information on necessary levels of health and safety training for those working on the project as well as arrangements for project-specific awareness training and refresher training such as 'toolbox' talks;

- [ ] sets out arrangements for monitoring compliance with health and safety law;

- [ ] includes, where appropriate, site health and safety rules and relevant health and safety standards, particularly where standards above the minimum statutory requirement are requested by the client.

The plan should be developed as far as possible before construction work starts and then reviewed and developed as necessary to account for changing project circumstances.

What is the health and safety file?
This is a record of information for the client/end user which the planning supervisor ensures is developed as the project progresses, and is produced at the end of the project before being passed to the client. It gives details of health and safety risks that will have to be managed during maintenance, repair or renovation. Contractors should pass information on these matters which becomes available during the construction phase to the planning supervisor for inclusion within the file.

The client should make the file available to those who will work on any future design, construction, maintenance or demolition of the structure.

What do the CDM Regulations require?

The client

Clients should:

- [ ] appoint a planning supervisor and principal contractor for each project;

- [ ] take reasonable steps to satisfy themselves that the planning supervisor, the principal contractor, project designers and any contractors they appoint directly, are competent and adequately resourced to deal with health and safety problems associated with the project;

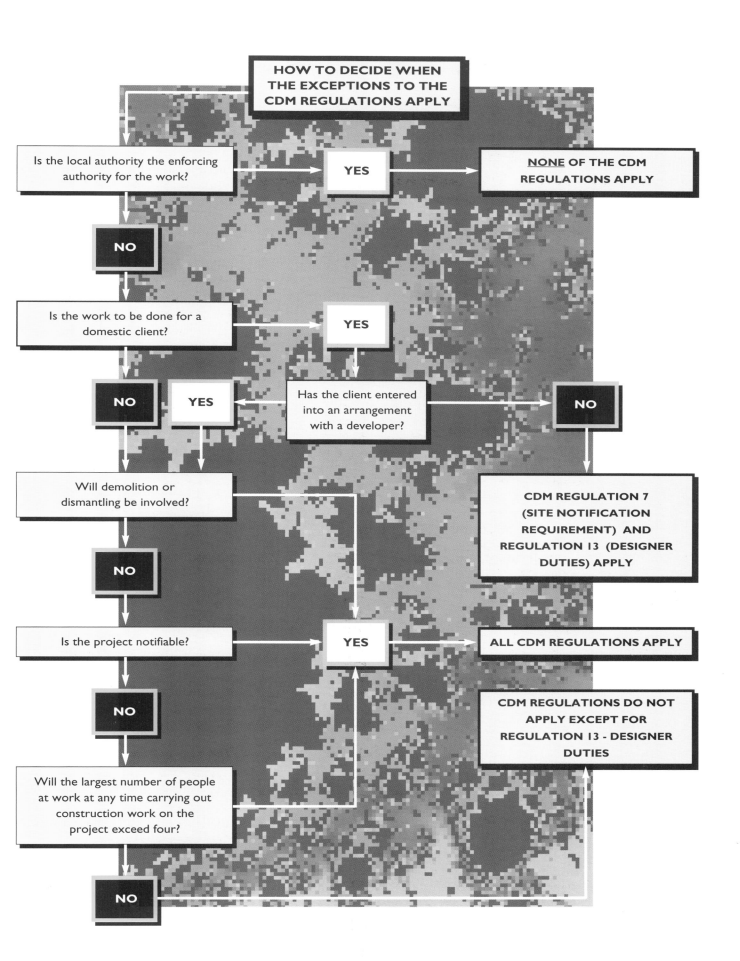

HOW TO DECIDE WHEN THE EXCEPTIONS TO THE CDM REGULATIONS APPLY

Is the local authority the enforcing authority for the work?

YES → **NONE OF THE CDM REGULATIONS APPLY**

NO

Is the work to be done for a domestic client?

YES

NO / YES

Has the client entered into an arrangement with a developer?

NO

Will demolition or dismantling be involved?

NO

CDM REGULATION 7 (SITE NOTIFICATION REQUIREMENT) AND REGULATION 13 (DESIGNER DUTIES) APPLY

Is the project notifiable?

YES → **ALL CDM REGULATIONS APPLY**

NO

CDM REGULATIONS DO NOT APPLY EXCEPT FOR REGULATION 13 - DESIGNER DUTIES

Will the largest number of people at work at any time carrying out construction work on the project exceed four?

NO

☐ pass on relevant information reasonably available to them about health and safety matters relating to the project to those planning the project. If a health and safety file is already available, relevant sections of this should be provided;

☐ ensure that construction work does not start unless a suitable construction phase health and safety plan has been prepared.

Clients may appoint agents to act on their behalf, but before doing so they should make reasonable enquiries to satisfy themselves that the agent is competent to fulfil the client's duties.

The designer

The term designer is widely drawn and includes everyone preparing drawings and specifications for the project. Designers include architects, structural engineers and surveyors. Before preparing any design the designer should ensure that the client has been made aware of the duties placed upon clients by the CDM Regulations and the guidance in the ACoP[57].

Designers should ensure that when they design for construction work they consider foreseeable health and safety risks during construction and eventual maintenance and cleaning of the structure in the balance with other design considerations such as aesthetics and cost. They should apply the hierarchy of risk control. This means designers need to identify the hazards from constructing the design and where possible avoid them. If the hazards cannot be avoided the design should minimise the risks and provide information about the risks that remain.

The design should describe any matters which require particular attention by a contractor. Enough information should be provided to alert contractors and others to matters which they could not reasonably be expected to know about.

The designer should also consider in the same way how the structure can be maintained and repaired safely once built. Designers should do this when they develop almost any design, including design work for projects where the appointment of a planning supervisor or principal

contractor is not required by the CDM Regulations.

Examples of what designers can do to improve health and safety are:

▪ designing for non-fragile rooflights instead of fragile ones;

▪ avoiding the need for chasing for cable runs (a job which inevitably exposes workers to high dust and noise levels) by embedding conduit within the wall finish;

▪ when designing foundations in contaminated land, specifying a driven pile foundation (which does not bring contaminated material to the surface) instead of bored piles;

▪ not specifying sharp-edged fishtail brick ties;

▪ avoiding concrete blocks weighing more than 20 kg which are difficult to lift and are likely to lead to long-term back injury to blocklayers.

Designers should co-operate with the planning supervisor and other designers on health and safety matters and supply relevant information. Information can be passed via the planning supervisor; if a planning supervisor does not need to be appointed, it should be supplied as part of the design information provided to the contractors. The information could include:

▪ the principles of the design relevant to the health and safety of those working on the project (eg erection sequences which must be followed to ensure stability);

▪ description of special requirements for safe working, (eg particular permit-to-work requirements);

▪ any special assumptions the designer has made about working practices (eg the site will have been levelled before the start of structural erection to allow the safe use of MEWPs for access for erectors).

If your firm provides any sort of design service to the client, or others, or designs

temporary works, this part of the Regulations will apply to you and you need to know more. Detailed advice for designers can be found in *Designing for health and safety in construction*[58], *Information on site safety for designers of smaller building projects*[59] and the ACoP[57].

The planning supervisor

The planning supervisor is appointed by the client. The planning supervisor may be an individual or a company and can be part of the client's organisation, one of the project designers, or some other person, partnership or organisation. Their job is to:

☐ co-ordinate health and safety during the design and planning phase of the project;

☐ ensure that the pre-tender stage health and safety plan for the project is produced in time for it to be provided to bidding contractors as part of the selection process;

☐ give advice about health and safety competence and resources needed for the project;

☐ ensure that written notification of the project is sent to HSE (see page 7); and

☐ collect information for inclusion in and ensure the preparation of the health and safety file which is passed to the client on completion of the contract.

The principal contractor

The principal contractor is appointed by the client to plan, manage and control health and safety during the construction phase of the project.

Site work should not start until the principal contractor has developed a construction phase health and safety plan based upon information provided by the planning supervisor. The client has to check that this has been done. The plan may need to be developed during the construction phase to take account of changing conditions on site as work progresses or the design changes.

When planning the job the principal contractor needs to identify the hazards and assess the risks of the work. To do this properly information, including health and safety method statements and risk assessments, may be needed from other contractors who will be working at the site.

When risks arise because of potential interactions between contractors (eg site transport matters) or a number of contractors are exposed to a common risk (eg from the site electrical distribution system), or use the same equipment, the principal contractor should take a positive role in ensuring the general principles of risk prevention and control are applied.

The principal contractor's health and safety plan should take account of the general issues in Section 1, the specific hazards and risk control measures in Section 2 and the general principles of risk assessment in Section 3.

Contractors

These are the firms or self-employed people working at the site. They should assist the principal contractor in achieving safe and healthy site conditions. They should co-operate with other contractors working on the site and provide health and safety information (including risk assessments - see pages 49 - 50) to the principal contractor.

REMEMBER:

If you manage or work on sites where CDM applies or act as a planning supervisor you may need to know more - see the Approved Code of Practice[57] which accompanies the Regulations and *A guide to managing health and safety in construction*[60] which tells you what to do to comply with the Regulations.

If you work on larger sites where CDM applies you should ask about the project health and safety plan before you start work. Your employees need to know what it says that affects them. Check that your working methods fit in with the plan and with site rules. If they do not, tell the principal contractor.

If any of your work requires you to do design work, even for temporary works, the Regulations apply to the design aspect, even if the Regulations do not require the appointment of a planning supervisor or a principal contractor. See *Designing for health and safety in construction*[58] for more information.

Health and safety competence

Everyone letting or subletting contracts should satisfy themselves that the people who will do the work (designers and contractors) are competent and resourced. They need to satisfy themselves that those who are to do the work are:

- competent in relevant health and safety issues; and

- intend to allocate adequate resources, including time, equipment and properly trained workers to do the job safely and without risks to health.

Assessing competence

If you are a client letting work, or a builder or contractor subletting work, asking questions about the issues in this section and Section 2 of the book will help you to assess competence. Decide in advance the competences needed and how they can be demonstrated. The pre-tender stage health and safety plan should act as a guide.

If you are tendering for work, being able to answer questions on these subjects will help you to demonstrate your competence and suitability for the job.

Employees' duties

Employees also have health and safety duties. They should:

- follow instructions given to them by their supervisors;

- co-operate with their employer on health and safety matters;

- follow the health and safety rules which apply to their particular job and to the site in general;

- use the health and safety equipment provided;

- report defects in equipment to their supervisor;

- take care of their own health and safety as well as that of their workmates and others who might be affected by their work.

Employees should be trained to know what to do and the work should be supervised and monitored to make sure that the information and training provided is relevant to the work situation and is applied effectively.

Employment and employees

Deciding on whether or not somebody is an employee, or is self-employed, can be complex in the construction industry. It is important to be sure of the employment status of people working on a site. It affects who has responsibility for some aspects of health and safety and the provision of safety equipment such as boots and hats.

Remember, just because someone pays their own tax and insurance, and holds a 714 or SC60, does not necessarily mean that they are self-employed. Deciding who is an employee depends on a range of issues. A person is more likely to be an employee when the following apply:

- they are paid an hourly rate;

- they are not allowed to subcontract work;

- they can be told by another (their employer) when, how and where they are to work;

- tools and materials are provided for them;

- the person has not entered into a contract for a fixed sum for a package of work.

However, these tests are not always certain and you may need to get legal advice to be sure of the situation. If you employ anyone you will need cover under the Employers Liability (Compulsory Insurance) Act 1969 and to display a current certificate.

Inspectors and the law

Health and safety laws which apply to your business are enforced by an inspector either from HSE or, where CDM does not apply, from the local council. Their job is to see how well you are dealing with your site hazards, especially the more serious ones which could lead to injuries or ill health. They may wish to investigate an accident or a complaint.

Inspectors do visit workplaces without notice but you are entitled to see their identification before letting them look around. Don't forget that they are there to give help and advice, particularly to smaller firms which may not have a lot of knowledge. When they do find problems they will try to deal with you in a reasonable and fair way. If you are not satisfied with the way you have been treated, take the matter up with the inspector's manager, whose name is on all letters from HSE. Any complaint about HSE inspectors will certainly be investigated, and you will be told

what is to be done to put things right if a fault is found.

Inspectors do have wide powers which include the right of entry to your premises, the right to talk to employees and safety representatives and to take photographs and samples. They are entitled to your co-operation and answers to questions.

If there is a problem they have the right to issue a notice requiring improvements to be made, or (where a risk of serious personal injury exists) one which stops a process or the use of dangerous equipment. If you receive an improvement or prohibition notice you have the right to appeal to an industrial tribunal.

Inspectors do have the power to prosecute a business or, under certain circumstances, an individual for breaking health and safety law, but they will take your attitude and safety record into account.

4 REFERENCES AND FURTHER INFORMATION

References (* = free)

1 HSE *The costs of accidents at work* HS(G)96 HSE Books 1993 ISBN 0 7176 0439 X

2 HSE *Provision of toilet, washing and general welfare arrangements at small fixed sites* Construction Information Sheet No 18 HSE Books 1991*

3 HSE *First-aid in your workplace - your questions answered* IND(G)3L(rev) HSE Books 1990*

4 HSE *Safe use of ladders, step-ladders and trestles* GS 31 HSE Books 1984 ISBN 0 11 883594 7

5 HSE *Tower scaffolds* GS 42 HSE Books 1987 ISBN 0 11 883941 1

6 HSE *Section A: Record of inspections of scaffolding* Form F 91 Part 1 HSE Books ISBN 0 7176 0437 3

7 HSE *General access scaffolds* GS 15 HSE Books 1982 ISBN 0 11 883545 9

8 HSE *Safe erection of structures: Part 1: Initial planning and design* GS 28/1 HSE Books 1984 ISBN 0 11 883584 X

9 HSE *Safe erection of structures: Part 2: Site management and procedures* GS 28/2 HSE Books 1985 ISBN 0 11 883605 6

10 HSE *Safe erection of structures: Part 3: Working places and access* GS 28/3 HSE Books 1986 ISBN 0 11 883530 0

11 HSE *Safe erection of structures: Part 4: Legislation and training* GS 28/4 HSE Books 1986 ISBN 0 11 883531 9

12 HSE *Health and safety in demolition work: Part 1: Preparation and planning* GS 29/1 (rev) HSE Books 1988 ISBN 0 11 885405 4

13 HSE *Health and safety in demolition work: Part 3: Techniques* GS 29/3 HSE Books 1984 ISBN 0 11 883609 9

14 HSE *Health and safety in demolition work: Part 4: Health hazards* GS 29/4 HSE Books 1985 ISBN 0 11 883604 8

15 HSC *Handling building blocks* Construction Information Sheet No 37 HSE Books 1993*

16 HSE *Manual handling* Manual Handling Operations Regulations 1992 Guidance on Regulations L 23 HSE Books 1992 ISBN 0 7176 0411 X

17 HSE *Getting to grips with manual handling: A short guide for employers* IND(G)143L HSE Books 1993*

18 HSE *Manual handling: Solutions you can handle* HS(G)115 HSE Books 1994 ISBN 0 7176 0693 7

19 HSE *Construction goods hoists* Construction Information Sheet No 13 HSE Books 1988*

20 HSE *Safety at rack and pinion hoists* PM 24 HSE Books 1981 ISBN 0 11 883398 7

21 HSE *Inclined hoists used in building and construction work* PM 63 HSE Books 1987 ISBN 0 11 883945 4

22 British Standards Institution *Code of Practice for safe use of cranes: Part 1: General* BS 7121: Part 1: 1989 *Part 2: Inspection, testing and examinations* BS 7121: Part 2: 1991

23 HSE *Avoiding danger from underground services* HS(G)47 HSE Books 1989 ISBN 0 7176 0435 7

24 HSE *Confined spaces* Construction Information Sheet No 15 HSE Books 1991*

25 HSE *Entry into confined spaces* GS 5 (rev) HSE Books 1994 ISBN 0 7176 0787 9

26 HSE *Health risk management* HS(G)137 HSE Books 1995 ISBN 0 7176 0905 7

27 HSE *Protection of workers and the general public during the development of contaminated land* HS(G)66 HSE Books 1991 ISBN 0 11 885657 X

28 HSE *Personal protective equipment at work* Personal Protective Equipment at Work Regulations 1992 Guidance on Regulations L 25 HSE Books 1992 ISBN 0 7176 0415 2

29 HSC *Protective clothing and footwear in the construction industry* IAC(L)16 HSE Books 1991*

30 HSE *PPE: Principles, duties and responsibilities* Construction Sheet No 28 HSE Books 1993*

31 HSE *PPE: Head protection* Construction Sheet No 29 HSE Books 1993*

32 HSE *PPE: Hearing protection* Construction Sheet No 30 HSE Books 1993*

33 HSE *PPE: Eye and face protection* Construction Sheet No 31 HSE Books 1993*

34 HSE *PPE: Respiratory protective equipment* Construction Sheet No 32 HSE Books 1993*

35 HSE *PPE: General and specialist clothing* Construction Sheet No 33 HSE Books 1993*

36 HSE *PPE: Gloves* Construction Sheet No 34 HSE Books 1993*

37 HSE *PPE: Safety footwear* Construction Sheet No 35 HSE Books 1993*

38 HSE *A step-by-step guide to COSHH assessment* HS(G)97 HSE Books 1993 ISBN 0 11 886379 7

39 HSE *Asbestos dust: The hidden killer* HSE Books 1995 IND(G)187L*

40 HSE *Asbestos alert* IND(G)188P HSE Books 1995*

41 HSC *The control of asbestos at work* Control of Asbestos at Work Regulations 1987 Approved code of practice L 27 HSE Books 1993 ISBN 0 11 882037 0

42 HSC *Work with asbestos insulation, asbestos coating and asbestos insulating board* Control of Asbestos at Work Regulations 1987 Approved code of practice L 28 HSE Books 1993 ISBN 0 11 882038 9

43 HSE *Noise in construction* IND(G)127L (rev) HSE Books 1993*

44 HSE *Dust and noise in the construction process* CRR 73 HSE Books 1995 (available later this year) ISBN 0 7176 0768 2

45 HSE *Hand-arm vibration* HS(G)88 HSE Books 1994 ISBN 0 7176 0743 7

46 HSE *Construction (Head Protection) Regulations 1989* Guidance on Regulations HSE Books 1990 ISBN 0 11 885503 4

47 HSE *Reversing vehicles* IND(G)148L HSE Books 1993*

48 HSE *Steering towards safety* IND(G)192L HSE Books 1995 (available later this year)*

49 HSE *Avoidance of danger from overhead lines* GS 6 (rev) HSE Books 1991 ISBN 0 11 885668 5

50 Building Employers Confederation, The Loss Prevention Council, National Contractors Group *Fire prevention on construction sites* The joint code of practice on the protection from fire of construction sites and buildings undergoing renovation 1992 ISBN 0 902167 20-0

Available from: BEC, Federation House, 2309 Coventry Road, Sheldon, Birmingham B26 3PL. Tel: 0121 742 0824.

51 Fire Protection Association *Construction sites fire prevention checklist* A guide for insurers, surveyors and construction industry professionals 1994 ISBN 0 902167 75-8

Available from: Fire Protection Association, 140 Aldersgate Street, London EC1A 4HX. Tel: 0171 606 3757.

52 Department of Transport *Safety at street works and road works* Code of practice HMSO 1992 ISBN 0 11 551144 X

53 HSE *Accidents to children on construction sites* GS 7 (rev) HSE Books 1989 ISBN 0 11 885416 X

54 HSC *Health and Safety at Work etc Act 1974: The Act outlined* HSC 2 HSE Books 1975*

55 HSC *Health and Safety at Work etc Act 1974: Advice to employers* HSC 3 HSE Books 1975*

56 HSE *Selecting a health and safety consultancy* IND(G)133L HSE Books 1992*

57 HSC *Managing construction for health and safety* Construction (Design and Management) Regulations 1994 Approved Code of Practice L 54 HSE Books 1995 ISBN 0 7176 0792 5

58 HSC *Designing for health and safety in construction: A guide for designers on the Construction (Design and Management) Regulations 1994* HSE Books 1995 ISBN 0 7176 0807 7

59 HSE *Information on site safety for designers of smaller building projects* CRR 72 ISBN 0 7176 0777 1

60 HSC *A guide to managing health and safety in construction* HSE Books 1995 ISBN 0 7176 0755 0

Site Safe News is published twice a year. If you would like to receive copies regularly, write to Sir Robert Jones Memorial Workshops, Units 3 and 5-9, Grain Industrial Estate, Harlow Street, Liverpool L8 4XY (tel: 0151 709 1354/5/6).

How to obtain publications

HSE priced and free publications are available by mail order from:

HSE Books, PO Box 1999,
Sudbury, Suffolk CO10 6FS
Tel: 01787 881165; Fax: 01787 313995.

HSE priced publications are available from all good booksellers. HSE information sheets are available from the Construction National Interest Group, HSE, 1 Long Lane, London SE1 4PG. Tel: 0171 407 8911 and your local HSE area office.

Further information

Other enquiries should be directed to HSE's Information Centre, Broad Lane, Sheffield S3 7HQ. Tel: 0114 2892345; Fax: 0114 2892333.

Area office addresses

South West
Inter City House, Mitchell Lane,
Victoria Street, Bristol BS1 6AN
Tel: 01179 290681 Fax: 01179 262998

South
Priestley House, Priestley Road
Basingstoke RG24 9NW
Tel: 01256 404000 Fax: 01256 404100

South East
3 East Grinstead House
London Road, East Grinstead RH19 1RR
Tel: 01342 326922 Fax: 01342 312917

London North
Maritime House, Linton Road
Barking IG11 8HF
Tel: 0181 594 5522 Fax: 0181 591 5183

London South
(all construction in London covered by this office)
1 Long Lane, London SE1 4PG
Tel: 0171 407 8911 Fax: 0171 403 7058

East Anglia
39 Baddow Road, Chelmsford CM2 0HL
Tel: 01245 284661 Fax: 01245 252633

Northern Home Counties
14 Cardiff Road, Luton LU1 1PP
Tel: 01582 444200 Fax: 01582 444320

East Midlands
5th Floor, Belgrave House, 1 Greyfriars
Northampton NN1 2BS
Tel: 01604 21233 Fax: 01604 30460

West Midlands
McLaren Building, 35 Dale End
Birmingham B4 7NP
Tel: 0121 609 5200 Fax: 0121 609 5349

Wales
Brunel House, 2 Fitzalan Road
Cardiff CF2 1SH
Tel: 01222 473777 Fax: 01222 473642

Marches
The Marches House, Midway
Newcastle-under-Lyme ST5 1DT
Tel: 01782 717181 Fax: 01782 620612

North Midlands
Birkbeck House, Trinity Square
Nottingham NG1 4AU
Tel: 01159 9470712 Fax: 01159 9411577

South Yorkshire and Humberside
Sovereign House, 110 Queen Street
Sheffield S1 2ES
Tel: 01142 739081 Fax: 01142 755746

West and North Yorkshire
8 St Paul's Street, Leeds LS1 2LE
Tel: 01132 446191 Fax: 01132 450626

Greater Manchester
Quay House, Quay Street,
Manchester M3 3JB
Tel: 0161 831 7111 Fax: 0161 831 7169

Merseyside
The Triad, Stanley Road, Bootle L20 3PG
Tel: 0151 922 7211 Fax: 0151 922 5031

North West
Victoria House, Ormskirk Road
Preston PR1 1HH
Tel: 01772 259321 Fax: 01772 821807

North East
Arden House, Regent Centre
Gosforth, Newcastle-upon-Tyne NE3 3JN
Tel: 0191 284 8448 Fax: 0191 285 9682

Scotland East
Belford House, 59 Belford Road
Edinburgh EH4 3UE
Tel: 0131 247 2000 Fax: 0131 247 2121

Scotland West
375 West George Street
Glasgow G2 4LW
Tel: 0141 275 3000 Fax: 0141 275 3100

Printed and published by the Health and Safety Executive
C200   6/95

Questionnaire

Health and safety for small construction sites

To help us assess this publication, will you please complete and return this questionnaire to the address overleaf. Postage is free.

We may wish to contact a sample of respondents with a fuller survey in future. If you do not wish to be contacted again please tick this box ☐

Mr, Mrs, Ms, Dr, Other _____ Initials _____ Surname _____

Position _____

Name of business _____

Address _____

_____

Postcode _____ Telephone _____ Fax _____

Size of business? (Number of employees)
Fewer than 5 ☐ 5 - 10 ☐ 10 - 20 ☐ 20 - 50 ☐ 50 - 100 ☐ 100 - 250 ☐ Over 250 ☐ Self employed ☐

What is your main activity?
General building ☐	*Roofing* ☐	*Painting* ☐
Joinery ☐	*Bricklaying* ☐	*Plumbing* ☐
Other (please specify) ☐		

How did you hear about this publication?
Advertisement ☐	*HSE Inspector* ☐	*Trade Association* ☐
HSC Newsletter/News Bulletin ☐	*HSE Catalogue* ☐	*Mailshot* ☐
Local authority ☐	*Informal business contact* ☐	*Other (please specify)* ☐

Did you find the publication:
clear and straightforward?			*difficult to understand?*
1	2	3	4

Was the publication:
too technical?			*not technical enough?*
1	2	3	4

Was the publication:
well presented?			*poorly presented?*
1	2	3	4

Do you feel that the publication represents:
very good value?			*poor value for money?*
1	2	3	4

Was the publication helpful to you in identifying the health and safety risks associated with the work you do:
very useful			*not useful?*
1	2	3	4

Was the advice in the publication useful to you in identifying ways of controlling health and safety risks associated with your work:
very useful			*not useful?*
1	2	3	4

Did the publication help you to understand your responsibilities for health and safety:
very well	*well*	*a little*	*not at all?*
1	2	3	4

How much of the advice was relevant to the work you do:
all	*most*	*some*	*none?*
1	2	3	4

Have you bought any of the other titles in the CDM guidance package? Y N

Any other comments _____

Thank you for taking the time to answer these questions

513

BUSINESS REPLY SERVICE
Licence No. LV 5189

Health and Safety Executive
Room 303, Daniel House
Stanley Precinct
Bootle
Merseyside L20 3QY

FIRST FOLD

SECOND FOLD

THIRD FOLD

2

A

B

Tuck A into B to form envelope
Please do not staple or glue